AF738396

Emerging Trends in Mechatronics

Series Editor

Aydin Azizi, Oxford, UK

Aydin Azizi
Editor

Control Engineering in Mechatronics

Springer

Editor
Aydin Azizi
School of Engineering, Computing
and Mathematics
Oxford Brookes University, Wheatley
Campus
Oxford, UK

ISSN 2731-4855 ISSN 2731-4863 (electronic)
Emerging Trends in Mechatronics
ISBN 978-981-16-7774-8 ISBN 978-981-16-7775-5 (eBook)
https://doi.org/10.1007/978-981-16-7775-5

© The Editor(s) (if applicable) and The Author(s), under exclusive license to Springer Nature Singapore Pte Ltd. 2023
This work is subject to copyright. All rights are solely and exclusively licensed by the Publisher, whether the whole or part of the material is concerned, specifically the rights of translation, reprinting, reuse of illustrations, recitation, broadcasting, reproduction on microfilms or in any other physical way, and transmission or information storage and retrieval, electronic adaptation, computer software, or by similar or dissimilar methodology now known or hereafter developed.
The use of general descriptive names, registered names, trademarks, service marks, etc. in this publication does not imply, even in the absence of a specific statement, that such names are exempt from the relevant protective laws and regulations and therefore free for general use.
The publisher, the authors, and the editors are safe to assume that the advice and information in this book are believed to be true and accurate at the date of publication. Neither the publisher nor the authors or the editors give a warranty, expressed or implied, with respect to the material contained herein or for any errors or omissions that may have been made. The publisher remains neutral with regard to jurisdictional claims in published maps and institutional affiliations.

This Springer imprint is published by the registered company Springer Nature Singapore Pte Ltd.
The registered company address is: 152 Beach Road, #21-01/04 Gateway East, Singapore 189721, Singapore

Preface

In today's fast-paced world, the integration of various engineering fields is essential to develop innovative and advanced technologies. The emerging field of mechatronics and control systems engineering is a perfect example of this trans-disciplinary approach, where the principles of mechanical, electrical, and computer engineering merge to design complex systems.

The book, *Control Engineering in Mechatronics*, is an attempt to provide an in-depth understanding of the fundamental scientific principles and technologies used in the design of modern computer-controlled machines and processes. The book emphasizes the synergies in the design process and explores the challenges and opportunities for integrating diverse engineering disciplines.

The book consists of six chapters that cover a wide range of topics related to mechatronics and control systems engineering. In Chap. Agent-Based Control System for SMEs—Industry 4.0 Adoption with Lean Six Sigma Framework, the book presents an agent-based control system for SMEs (Small and Medium-sized Enterprises), and it explains how the Lean Six Sigma Framework can help in the adoption of Industry 4.0.

Chapter Path Tracking Control of Bioflexible Probes Exposed to Uncertainties and Internal Tissues Disturbances with Unknown Upper Bonds Using Robust-Adaptive Sliding Mode Control focuses on the path tracking control of bioflexible probes, which are exposed to uncertainties and internal tissue disturbances with unknown upper bounds. The book explores how robust-adaptive sliding mode control can be used to address these challenges.

In Chap. Applied Mechatronics: A Case Study on Mathematical Modelling and Experimental Analysis of the Second Life Batteries, the book presents a case study on mathematical modeling and economic analysis of second-life batteries, which is an emerging field in mechatronics. The chapter discusses how these batteries can be reused in different applications and how mathematical modeling can help in the economic analysis of these batteries.

Chapter Embracing Digital Technologies in the Pharmaceutical Industry focuses on the pharmaceutical industry, and the book explores the opportunities and challenges of embracing digital technologies in the industry. The chapter presents examples of how digital technologies can be used in the pharmaceutical industry to improve efficiency, reduce costs, and enhance quality.

Chapter Mathematical Modeling and Hybrid Adaptive-Fuzzy Control of HIV/AIDS Infection delves into the hybrid adaptive-fuzzy control of HIV/AIDS infection, and it explains how mathematical modeling can be used to develop a control system for the disease. The chapter explores the challenges and opportunities of using a hybrid adaptive-fuzzy control system for HIV/AIDS infection.

In the final chapter, the book discusses the impact of communication network characteristics on cyber-physical systems. The chapter presents an overview of the different communication network types and explores their impact on cyber-physical systems.

Overall, the book *Control Engineering in Mechatronics* is an excellent resource for professionals, engineers, researchers, and students who want to gain a comprehensive understanding of the trans-disciplinary field of mechatronics and control systems engineering.

Oxford, UK Aydin Azizi

Contents

About the Editor

Dr. Aydin Azizi holds a Ph.D. degree in Mechanical Engineering-Mechatronics, an M.Sc. in Mechatronics and a B.Sc. in Mechanical Engineering-Heat & Fluids. Certified as the Fellow of the Higher Education Academy (FHEA), official instructor for the Siemens Mechatronic Certification Program (SMSCP) and the editor of the book series Emerging Trends in Mechatronics publishing by Springer Nature Group, he currently serves as a Senior Lecturer at the Oxford Brookes University. His current research focuses on investigating and developing novel techniques to model, control and optimize complex systems. Dr. Azizi's areas of expertise include Control & Automation, Artificial Intelligence and Simulation Techniques. Dr. Azizi is the recipient of the National Research Award of Oman for his AI-focused research, DELL EMC's "Envision the Future" competition award in IoT for "Automated Irrigation System", and 'Exceptional Talent' recognition by the British Royal Academy of Engineering.

Agent-Based Control System for SMEs—Industry 4.0 Adoption with Lean Six Sigma Framework

Poorya Ghafoorpoor Yazdi, Aydin Azizi, and Majid Hashemipour

Abstract Small and Medium-sized Enterprises (SMEs) play a vital role in the world economic structure due to their significant contribution to production, exports and employment. However, there are various financial, marketing and production issues associated with SMEs. This is mainly due to weak traditional manufacturing systems and inflexible control architectures to respond to various market needs. In order to survive, SMEs must be able to overcome the rapid change of the markets and the diverse demands of customers. This involves achieving and maintaining high levels of productivity and the capability to respond rapidly and flexibly in a short lead time. The Industry 4.0 is a current manufacturing trend which improves efficiency, flexibility and agility, and increases the profitability of enterprises by offering different manufacturing paradigms. However, SMEs leaders have doubted the benefits of Industry 4.0 for implementation in their manufacturing system. One of the primary design principles of Industry 4.0 is "Decentralized Decisions" which potentially can address the problem of traditional control architecture if implemented. Therefore, this research was set out to implement "Decentralized Decisions" to facilitate the Industry 4.0 adoption and improve the efficiency of SMEs. Consequently, a distributed control system was required which was achieved by developing an agent-based control architecture with a Master–Slave mechanism. Lean Six Sigma (LSS) approach was utilized to recognize the limitations, assess, and maximise the system performance after implementing the developed control architecture. It was achieved by measuring the system production time using a time study technique

P. Ghafoorpoor Yazdi (✉)
Department of Design, Production and Management, University of Twente, Enschede, The Netherlands
e-mail: p.ghafoorpooryazdi@utwente.nl

A. Azizi
School of Engineering, Computing and Mathematics, Oxford Brookes University, Wheatley Campus, Oxford OX33 1HX, UK
e-mail: Aydin.azizi@brookes.ac.uk

M. Hashemipour
Faculty of Engineering, Cyprus International University, Nicosia 99258, North Cyprus
e-mail: majid.hashemipour@emu.edu.tr

© The Author(s), under exclusive license to Springer Nature Singapore Pte Ltd. 2023
A. Azizi (ed.), *Control Engineering in Mechatronics*, Emerging Trends in Mechatronics,
https://doi.org/10.1007/978-981-16-7775-5_1

that is used in performance evaluation which is based on Overall Equipment Effectiveness (OEE). A series of solutions were obtained and applied to a system simulation model to assess their influence on maximizing the performance. Since the OEE calculation is based on production time which is proportional to the distance between the resources and speed, the corresponding solutions were chosen accordingly. The behaviour of the resources in the system was different for each solution. Therefore, the solutions were prioritized based on their influence on OEE percentage. The OEE percentage improvements varied from 1 to 15% between the resources. It was observed that considering the highest solution priority for each resource results in maximum system performance. The target system for this research shared the characteristics and features of a SME and the results indicated that implementing the agent-based control architecture along with LSS improved the performance. Implementing both techniques provides a significant step towards successful SME adoption of Industry 4.0 and improves their response to the challenging market.

Keywords Small and medium sized enterprises · Industry 4.0 · Agent-based control · Lean six sigma · Time study · Overall equipment effectiveness · Simulation manufacturing performance measurement

List of Symbols or List of Abbreviations

SME	Small and Medium Sized Enterprises
OEE	Overall Equipment Effectiveness
LSS	Lean Six Sigma
AGV	Automated Guided Vehicle
MHESA	Material Handling Equipment Selection Advisor
MCDM	Multi Criteria Decision-Making
MTM	Methods-Time Measurement
MOST	Maynard operation sequence technique
LMS	Lean Manufacturing System
VA	Value Added
NVA	Non-Value Added
SVA	Semi-Value Added
FPS	Frame Per Second
IO	Input-Output
FS	Functional Specification
p_1	Production Time
p_2	Distance Between the Stations or Segments
p_3	Speed of the Motion
T_i	Idle Time
T_b	Busy time
C_1	Coefficient of Idle Time
C_2	Coefficient of Busy Time

1 Introduction

1.1 Background

Nowadays, business with increasing globalization and rapid technological changes forced enterprises to move their production abroad by considering various features of customer satisfaction. This Transformation requires more firms known as small and medium-sized enterprises (SMEs). SMEs are an important source of innovation, new products, and new services all over the world [1]. As a result, the category of SMEs has become increasingly critical to the previously mentioned growth [2]. SME firms must produce admirable products at a low cost, with quick market delivery and appropriate quality. However, most SMEs are limited in terms of knowledge and technology, and they have limited resources [3]. The strategy of meeting a wide variety of demands has increased the complexity, dynamic, and demanding nature of SMEs manufacturing systems [4]. On the other hand, the design of such manufacturing systems requires a suitable control architecture, a high level of expertise, and deliberate decisions to ensure that the system can successfully meet the demands of a constantly changing market [5]. One possible solution for SMEs is to use intelligent technology (IT) and Web-based enterprise management system (WBEMS) to provide more product options, minimise the cost of testing, improve business processes in the production and quality management cycle and provide better supply chain management services, thereby reducing the pressure on SMEs manufacturing systems [5]. Amongst the emerging strategies for coping with the challenges of manufacturing SMEs have to meet new management, engineering and IT issues in their endeavour to build and maintain a competitive advantage in the market.

Additionally, it is difficult and extremely risky for SMEs to implement these novel control architectures and manufacturing system types without first validating and verifying their potential benefits and drawbacks, as well as their effect on productivity [6]. As a result, modelling, simulation, and performance analysis are critical for the successful development and implementation of a new manufacturing system and control architecture for SMEs [7].

1.2 Industry 4.0

The term "Industry 4.0" refers to the most recent phase of the technologically-driven industrial revolution. Industry 4.0 is being led by previous industrial revolutions throughout the industry's history. Past Industrial Revolutions Although the term Industry 4.0 came into use in 2011, all of the industrial revolutions that began in the mid-eighteenth century were steps leading up to today's digital revolution. The first revolution began in the eighteenth century with the invention of mechanical production facilities that relied on steam and water power, such as spinning and weaving machines. These mechanical mills sparked growth in Europe and the United States and helped drive the Industrial Revolution. Human labour division and electrification were the catalysts for the second industrial revolution, which began in the 1870s. Another driving force for the second industrial revolution was the advent of mass production techniques and mechanical efficiency methods that were developed in the second half of the nineteenth century. The third revolution, dubbed the "digital revolution," began in the 1970s as a result of advancements in information technology, automation, and electronics development (as symbolized by the invention of the microchip), which led to increased automation and a vast expansion of worldwide telecommunications and computerization and the widespread adoption of digitized technology in nearly every aspect of daily life [8, 9].

Industry 4.0 is expected to improve the productivity of manufacturing industries, create and service new markets, empower industries that create a higher standard of living for people across the globe, and further support global economic growth. Also, it improves enterprises' performance by streamlining manufacturing processes, optimizing resource and material utilization, and streamlining supply chain and life cycle management [10]. The revolution has already had a major impact on the business world, dramatically increasing efficiencies and transforming how companies design, market, sell and distribute their products. Industry 4.0 implementation in businesses requires adherence to and adoption of certain principles driving the adoption of technologies for collecting, analyzing and interpreting a massive amount of data. In order to fulfil the ambitious objectives of Industry 4.0, it is critical for every organization in manufacturing and services to adopt modern technologies and follow these principles toward their effective implementation throughout the entire process flow. This list includes the principles of "interconnection," "information transparency," "technical assistance," and "decentralized decision-making". These principles are set to change the way manufacturing and services are done and how they are perceived. The primary strategy for implementing Industry 4.0 in SMEs is to enable intelligent communication between human and hardware resources, for the purpose of enhancing cost-effectiveness and productivity, enabling continuous improvement in products, processes and services [11]. Productivity within and outside of a company's physical boundaries can be improved by integrating intelligent communication between various components, including people, machines, material handling systems, and information technology (IT) tools. When implemented, this brings with it the ability to continuously optimize products, processes and services. This is especially true in small and medium-sized enterprises (SMEs) through the use of new technologies that allow machines to connect to one another as well as with people [12].

1.2.1 Industry 4.0 Principals

Interconnection, information transparency, decentralized decision-making, and technical assistance have been identified as the four guiding design principles for the implementation of Industry 4.0, as was discussed in the previous section. These principles will be elaborated upon in the following sections.

Interconnection

The Internet of Everything is formed by the interconnection of people, machines, devices, and sensors through the Internet of Things (IoT) and the Internet of People (IoP). Because they make it possible to access the internet almost anywhere, wireless communication technologies play a significant part in the expansion of human interaction. Interconnected people and things are able to share information through the Internet of Everything, which serves as the foundation for joint collaboration aimed at accomplishing shared objectives [12]. Within the IoE, there are three distinct ways in which individuals can work together: human–human collaboration, human–machine collaboration, and machine-machine collaboration [13]. The establishment of common communication standards is of the utmost significance for facilitating the linking of individuals, machines, devices, and sensors. These standards make it possible to combine modular machines from a variety of manufacturers in a flexible manner [14]. Because they are modular, the Smart Factories of Industry 4.0 are able to adapt flexibly to changing market demands as well as personalized orders. As the number of participants in the IoE grows, monetary and political interests will lever the number of harmful attacks on Industry 4.0 production facilities, which will consequently increase the need for cyber security [15].

Technical Assistance

As a result of Industry 4.0's technical assistance, humans are no longer required to calculate and solve problems manually, but rather with the aid of computing software and simulators that facilitate visualization. Technical assistance also refers to machines that perform physically exhausting or unpleasant tasks that workers would otherwise have to perform. When a product is used, technical support represents the connection between the product and the production system. All information collected by an intelligent product is returned to the production system. This provides users with guidance in the event of a product malfunction. In addition, it enables production systems to analyzes manifested errors and improve future products based on their findings. First, technical assistance is the capacity of cyber-physical systems to aid humans by aggregating and visualizing information in a way that makes making informed decisions and solving pressing problems in a timely manner simple and efficient. Work instruction software is a prime example of Technical Assistance used to assist individuals with visually interactive procedure guides. Second, technical assistance is the capacity of cyber-physical systems to physically assist humans by performing a variety of tasks that are unpleasant, too challenging, or unsafe for their human coworkers. Consider the current role of robots in factories and how they reduce workplace injuries. As a result of technical assistance, manufacturers are able to comprehend their data in a more efficient and effective manner, allowing them to focus on the more difficult and hazardous tasks on the shop floor.

Control and decision-making should be decentralized as a solution to make it easier to make changes in the production process, which will help the company meet the growing demand. Decentralized decision-making capability enables each definable segment of the SMEs production system to act autonomously in carrying out their assigned tasks. Decisions will be made in discrete locations throughout the system to maximize response time and flexibility while the system remains operational [11]. This characteristic results in intelligent production, which includes knowledge of manufacturing history and products, as well as the ability to actively steer products through the manufacturing process by instructing equipment to perform required manufacturing tasks [16].

1.3 Material Handling Systems in Intelligent Production Systems

The material handling system (MHS) is a critical component of deploying an intelligent manufacturing system. However, before enterprises can implement intelligent production in the context of Industry 4.0, certain requirements for MHS must be considered [17, 18]. In the majority of SMEs, the material handling process involves employees devoting significant time and effort to manually control the MHS, resulting in increased production costs and lead times [19]. However, the concept of intelligent material handling has been developed in conjunction with Industry 4.0 implementation, and it is all about leveraging new technologies to create more intelligent control systems capable of making better decisions [9, 20]. For the majority of businesses that still use traditional material handling systems, it would be risky and in some cases impossible to make any changes to convert to intelligent MHS. This transformation necessitates the development of a new control architecture in order to make the system intelligent. As a result, enterprises prefer to evaluate and validate any effect that the new control architecture may have on their existing MHS [21].

1.4 Types of Control Systems

Industry 4.0 adoption in enterprises in order to reap the benefits of an intelligent production system will necessitate some modifications to the enterprise's control system. In contrast to the centralized control systems that are prevalent today, control systems should be distributed across the entire shop floor or resource base [22].

1.4.1 Centralized Control System

A central control unit controls the operations of all the corresponding individual units in a centralized control system. The central controller is responsible for decision-making, and the operation of all corresponding control units is contingent upon it.

Individual controllers lack the ability to communicate with and collaborate with one another [23].

1.4.2 Decentralized Control System

In a decentralized system, decision-making is not delegated to a central control unit. This type of control system has a hierarchical structure that represents a combination of middle controllers (nodes) that control all external nodes and communicate with a central control unit. Each controller in a decentralized control system is controlled by those at the higher level and can control those at the lower level. As a result, a central controller can control the entire system [24].

1.4.3 Distributed Control System

In comparison to centralized and decentralized control systems, a distributed control system is distributed. In this type of system, each controller has equal authority and decision-making authority. Each controller in this class of control system is referred to as an agent [25]. Each agent is capable of communicating with others by adhering to certain protocols, principles, and architecture. The system's distinguishing feature is that if one of the agents is unable to perform a task, the other agent can. As a result, increasing the number of agents increases the system's reliability [26, 27].

1.4.4 Control Agents

Agent-based control systems have evolved into a critical component of manufacturing control systems. An agent is a dynamic object that possesses the capability to perform tasks and communicates with other agents via the organizational structure in order to collaborate on task completion [28]. Current manufacturing systems for SMEs are incapable of keeping up with product evolution and market changes. Additionally, SMEs must maintain a satisfactory level of performance outside of normal business hours. The agent paradigm is intended to overcome these obstacles by incorporating critical concepts such as automation and cooperation [29]. In this situation, agent technology emerges as the ideal candidate to take on the new challenge and enhance the capability of Industry 4.0 implementation [27].

1.4.5 Agent-Based Control System

This type of control system is referred to as a distributed control system, and it serves as a guide for developing a software system with the assistance of numerous existing agents [30]. Agent-based modelling (ABM) depicts a framework as a collection of autonomous decision-making agents. Each agent independently examines the system's state and makes decisions based on a set of rules and algorithms. Agents can exhibit a variety of behaviours that are appropriate for the correspondence system [31].

1.4.6 Master and Slaves Agents

When it comes to computer networking, devices typically operate in one of two modes: slave mode or master mode. In a network, the master device, also known as the principal device, is in charge of synchronizing communication throughout the network in accordance with specific protocols and principles. Because of this, the remaining devices are regarded as slave devices in the system. The slave devices, on the other hand, are permitted to connect to the master as well. In addition to being able to send data to any of its slaves, the master can also request feedback from those slaves. Because of this, slaves are only permitted to send and receive data to and from one another with Master confirmation and instructions [32].

1.5 Lean Six Sigma Approach for Performance Evaluation and Improvement of the Manufacturing System

Industry 4.0 implementation has significant and widely applicable effects or implications on businesses. There is still a dearth of research into these impacts and influences on businesses [33]. The most difficult impacts to investigate are those that result from the use of novel control architectures [34]. While the new control architecture is expected to be beneficial, it is also extremely challenging given the possibility of enterprise performance improvements [35]. The expectation is that the new control architecture will have a positive effect on the manufacturing system's performance. Thus, the rigours, risks, and constraints that enterprises may face following their implementation should be investigated.

Six Sigma is a collection of approaches and methodologies, including statistical tools and techniques, for investigating, detecting, and eliminating the defects and limitations mentioned previously, as well as increasing process efficiency [36]. Lean and Six Sigma (LSS) have recently become the most widely used method for enterprises to implement continuous improvement (CI) [36]. For any enterprise, particularly small and medium-sized enterprises, the primary goal is continuous improvement in operational quality and performance [37, 38]. Thus, implementing LSS in SME manufacturing processes improves the relationship between manufacturing processes and variability, thereby reducing defects. Additionally, LSS focuses on process improvement through a thorough and accurate examination of causal relationships via the collection and analysis of real data. The LSS process entails activities for measurement, improvement, and validation. Emerging methodologies and technologies that are required for enterprises to implement industry 4.0 have

Fig. 1 Lean six sigma

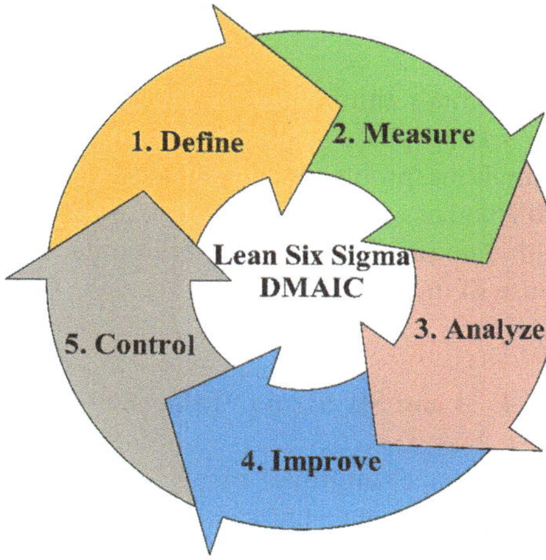

emphasized the importance of an LSS approach [13]. LSS follows the established cycle of Define, Measure, Analyze, Improve, and Control (DMAIC) [13, 14, 37]. (Fig. 1).

1.5.1 Lean Six Sigma "Define" Phase

The first aspect of Six Sigma to consider is the Define phase, which involves defining the Problem Statement. Refining the problem statement entails reducing the scope of the issues and constraints in order to increase focus. Six Sigma's "Define" phase identifies manufacturing system problems, limitations, and defects, as well as the enterprise's goals. The problem statement and objectives must be "Specific," "Measurable," "Achievable," "Realizable," and "Time-bound" [15].

1.5.2 Lean Six Sigma "Measure" Phase

Six Sigma's second phase involves quantifying the performance of the manufacturing system and associated processes, which will be used to identify areas for improvement. There are numerous factors that affect the performance of a manufacturing system, but the ones with the greatest influence should be identified and quantified. Data from the measured phase will be compared to data from the system's outcome following modification to determine whether the improvement was effective or if another improvement is required [38].

1.5.3 Lean Six Sigma "Analyze" Phase

Six Sigma's third phase is the analysis of the measured data that resulted from the second phase. This phase primarily identifies in detail the primary causes of limitations, problems, and defects. All processes in the manufacturing system, along with their associated tasks, will be analyzed. This analysis is conducted in conjunction with data on the target factors in order to determine which factor(s) is/are the root cause of the defect [38, 39].

1.5.4 Lean Six Sigma "Improve" Phase

Six Sigma's "Improve" phase identifies solutions and ideas for resolving or minimizing problems, constraints, and defects. The selected factors' values will be set using a variety of process variables. Adjusting process variables can optimize the selected factor, thereby improving performance [37, 39].

1.5.5 Lean Six Sigma "Control" Phase

The control phase of Six Sigma is the final stage, with the primary objective of sustaining the improvements made to the manufacturing system and its associated processes. Numerous tools will be used to determine which variables are the most effective at improving the performance of the selected factors [39].

1.6 Overall Equipment Effectiveness (OEE)

Considering the "Analyze" phase of Six Sigma approaches for Industry 4.0, determining the equipment's effectiveness is critical for achieving a low defect rate and increased productivity [14, 40, 41]. Overall Equipment Effectiveness (OEE) is a method for evaluating the performance of businesses, particularly small and medium-sized enterprises (SMEs) [42]. The OEE is defined as the difference between the valuable operating time and the loading time. The term "operation time" refers to the period of time during which the equipment produces acceptable products. Whereas loading time is the amount of time required for equipment to operate during a specified period [43].

Standard metrics are critical for evaluating an enterprise's performance and productivity. These metrics can be used to determine how the manufacturing system's performance can be improved [44]. Each metric can be used to assess various aspects of production performance, including efficiency, quality, flexibility, inventory management, and profitability. The overall equipment effectiveness (OEE) metric is one used to determine the percentage of truly productive time. It is made up of three components: availability, performance, and quality. Thus, conducting OEE assists in identifying and resolving inefficiencies and bottlenecks that may occur during the manufacturing process [44].

Additionally, OEE evaluations of manufacturing systems are not limited to determining manufacturing lead times. OEE analysis enables a systematic process for identifying common sources of productivity loss [45]. Additionally, OEE evaluation can help reduce costs, increase awareness, increase machine productivity, and extend the life of the equipment [46, 47].

Three factors must be evaluated in order to determine the OEE. "Availability," "Performance," and "Quality" are these factors. Equation (1) illustrates the relationship between these variables and the evaluation of OEE [45].

$$\text{OEE} = \text{Availability} \times \text{Performance} \times \text{Quality} \tag{1}$$

Availability is the ratio of run time to planned production time, and it takes into account availability loss, which is caused by production process stoppages. Unplanned stops can occur as a result of equipment failure or a shortage of materials, while planned stops can be a result of changeovers. The time remaining after subtracting all available losses from the total production time is referred to as the run time [44]. The following Eq. (2) can be used to determine availability:

$$\text{Availability} = \frac{\text{Run time}}{\text{Planned production time}} \tag{2}$$

If the availability is 100%, it can be concluded that no downtime occurred during the entire production period [49, 50].

The OEE's second factor is performance. The performance factor accounts for anything that contributes to the production process running slower than the maximum possible speed (including slow cycles and small stops) [51]. The following Eq. (3) will be used to determine performance:

$$\text{Performance} = \frac{(\text{ideal cycle time} \times \text{total count})}{\text{Run time}} \tag{3}$$

In this case, Ideal Cycle Time refers to the time required to produce one unit of product. The total count refers to the total number of products, including defective products [52].

Another critical factor in calculating OEE is quality. This factor takes into account all of the products produced during the manufacturing process. It also accounts for quality losses, which are products with defects that do not meet defined quality standards and must be reworked to achieve a desired level of quality. The quality factor used in OEE evaluation takes into account all manufactured components, regardless of whether they met quality standards. The following equation will be used to determine the quality:

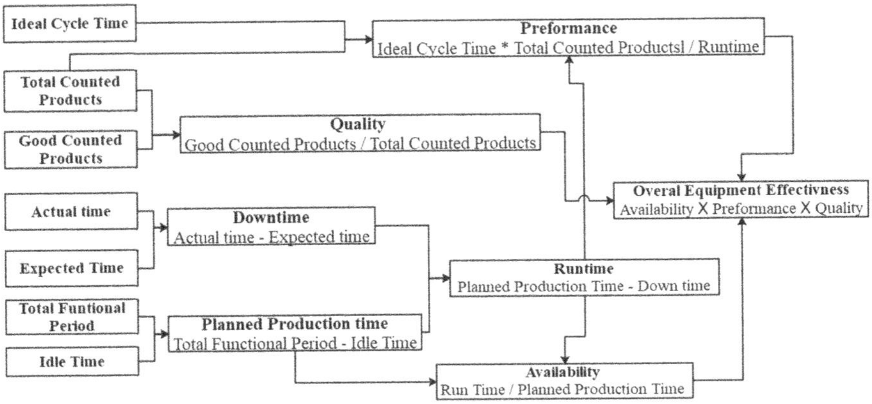

Fig. 2 OEE overall procedure

$$\text{Quality} = \frac{\text{good count}}{\text{total count}} \tag{4}$$

The good count is the number of manufactured products that meet quality standards, while the total count is the total number of manufactured parts [53]. The overall procedure for calculating the OEE is depicted in a flowchart (Fig. 2).

1.7 Lead Time and Time Study in OEE and LSS

Time is the most critical factor in analyzing the OEE across enterprises. To use the OEE standard to evaluate the system's performance, a comprehensive timing overview of the target system must be created [54]. A proper time study approach monitors the behaviour of the entire production process, as well as the resources and tasks that are associated with it. Conducting a time study enables the identification and elimination of inefficient time and bottlenecks that may occur during the process [55]. As a result, manufacturing lead time is a highly effective factor in the "Measure" phase of Six Sigma due to its critical role in OEE evaluation, and time study techniques are critical in the "Analyze" phase of Six Sigma.

Data for time studies should be entered in a consistent format. There are several time study techniques available for evaluating manufacturing performance. The most well-known techniques are Methods-Time Measurement (MTM) [56], Maynard operation sequence (MOST) [57] and Modular Arrangement of Predetermined Time (MODAPTS) [58]. To obtain an accurate time study result for the system, a combination of the above techniques eliminates the time study data's lack of visibility across the organization. The procedure may be facilitated by combining the aforementioned techniques with video studies. A time study tool can be used to import captured video of the manufacturing process. The video can be segmented according to available resources and tasks. It enables the rapid generation of time study results and enhances their traceability [59, 60].

1.8 The Simulation for OEE and LSS

Simulation is critical for evaluating, validating, and verifying any optimization idea or change to the manufacturing system's hardware, software, or layout design [61]. The operational performance of manufacturing systems can be determined and validated prior to implementing any changes to the simulation model on the real system [62]. Simulation is the behavioural replication of real-world processes or systems over time [63], whereas optimization is the process of determining the best element within a given definition domain in terms of certain criteria. In the last few years, simulation has become an increasingly effective tool for optimizing manufacturing systems [64, 65].

Given the definition of simulation and its significance in evaluating the performance of manufacturing systems, it is reasonable to conclude that simulation plays a significant role in the Six Sigma "Measure" phase. Thus, for any possible solution to improve the performance of manufacturing systems, simulation can be used to determine the factors that have the greatest effect on the manufacturing system's performance. However, prior to implementation, it is necessary to validate the identification of these factors and the manner in which they must be modified to improve the manufacturing system's performance. Due to the aforementioned characteristics of simulation, it is one of the most effective tools in the Six Sigma "Improve" phase.

Simultaneous with the implementation of industry 4.0 is the emergence of decision-making distribution between control devices. The fundamental goals of distributed decision-making are to comprehend, analyze, and optimize the operational conditions of systems [66]. While classical control architectures continue to play an important role in achieving operational goals, their applicability to modern industrial systems is limited. As Industry 4.0 is frequently used to refer to modern industrial systems that are more interconnected and responsive, the prospect of comprehensively understanding the influence of new control architecture on these system operations has become critical. Before implementing new control architectures, simulation is an effective tool for obtaining a broad and powerful overview of industrial system performance [67].

ARENA is a general-purpose discrete event simulation and automation software used in a variety of industries (supply chain, manufacturing, logistics, and military) to address a variety of business and industry challenges [68]. ARENA simulation can be used to reproduce various system configurations that have been subjected to optimization solutions, serving as a validation, diagnostic, and verification framework for the optimization module's proposed solutions [69–71].

1.9 The Objective of the Research

Due to competitive global market demands over the last decades, SMEs have been compelled to enhance their capabilities through integration into Industry 4.0. Industry 4.0 is a concept that SMEs can implement primarily through the use of new collaborative, agile, and responsive control architectures. A well-designed and implemented control architecture results in increased performance for SMEs, which is critical for manufacturing success on a global scale. However, SMEs face numerous uncertainties when implementing new control architectures. The purpose of this research is to develop a novel agent-based control architecture that enables the adoption of Industry 4.0 by SMEs through distributed control. It is critical that the proposed control architecture provide sufficient compelling reasons and solutions to assist SMEs in overcoming their fears of modifying their existing control system. The methodology identifies issues and constraints and makes recommendations for performance enhancements. The methodology is based on the Lean Six Sigma philosophy, with OEE serving as the performance measurement standard and an accurate time study technique used to identify the system's defects and limitations. Additionally, a simulation tool was used to deploy optimization solutions for resolving identified issues and defects prior to their implementation in the real system. Implementing the novel agent-based control architecture and methodology proposed in this research will result in improved manufacturing system performance and may enable SMEs to take the first and most critical step toward implementing Industry 4.0.

The following outline outlines the steps necessary to accomplish the study's objectives:

1. Choosing a system as the target system that incorporates the characteristics of SMEs
2. Inquiry into the target system regarding the following:
    - System functionality and the current state of affairs
    - The current control architecture and the control units that are available, as well as their functions
    - Availability of resources and associated task(s).
3. Using the following steps to create a novel agent-based control architecture:
    - Assigning control units to Master and Slave agents
    - Define the tasks assigned to each Slave agent (a slave for each Resource)
    - Assignment of tasks to the Master agent
    - Developing agent software for Master and Slave agents that includes the ability to make precise decisions
    - Developing protocols and principles for communication between the master and slaves
    - Developing the slaves' communication protocols

4. Using the Lean Six Sigma-DMAIC process to evaluate and improve the target manufacturing system's performance, including the developed agent-based control architecture.
5. Passing through the five critical phases of Six Sigma in the following order:
   - Define Phase:
     - Defining the target manufacturing system and its associated processes' problems and constraints
     - Identifying all of the target manufacturing system's available resources
     - Identifying the manufacturing system's production plan
   - Measure Phase:
     - Creating a detailed map of the target manufacturing system's processes
     - Creating a data collection strategy
     - Collecting and analyzing all necessary data
     - Validation of gathered data
   - Analyze Phase:
     - Analyzing the data gathered in the previous phase
     - Identifying issues with individual resources, processes, or the entire target manufacturing system
     - Identifying the Causal Factors for Each Detected Issue
     - Identifying the underlying cause of detected issues.
   - Improve Phase:
     - Identifying possible and potential solutions to the problem identified in the previous phase
     - Analyzing the solutions' failure modes
     - Identifying the desired improvements to the manufacturing system
     - Validation of the enhancements
   - Control Phase:
     - Implementing a statistical process control system to monitor the target manufacturing system both after and before the solutions are implemented.
6. Integrating the target manufacturing system with the developed agent-based control architecture in order to resolve fundamental issues ("Define" Phase of LSS)
7. Subdivide the target system into distinct sections, each with its own associated resource ("Define" Phase of LSS)
8. Choosing an appropriate time study technique and tool ("Measure" Phase of LSS)

9. Collecting time study data from each section as they perform their assigned tasks and compiling a comprehensive time study database ("Measure" Phase of LSS).
10. Accuracy validation of the time study data collected ("Measure" phase of LSS)
11. Analyzing the results of the time study in order to obtain the information necessary for performance evaluation of the target system and its associated resources. ("Measure" Phase of LSS)
12. Choosing an appropriate technique for evaluating performance such as Overall Equipment Effectiveness ("Analyze" Phase of LSS).
13. By incorporating the results of the time study into the OEE, we can perform performance evaluations on individual resources and the entire system ("Analyze" phase of LSS).
14. Identifying system problems and constraints through an analysis of time study and OEE results. ("Analyze" Phase of LSS)
15. Classification of identified issues into hardware, layout design, and control architecture ("Analyze" phase of LSS)
16. Determining the most effective factors that contribute to the existence of identified problems ("Analyze" Phase of LSS)
17. Making recommendations for potential solutions (optimization solutions) based on the identified effective factors. ("Improve" LSS Phase)
18. Developing the system's simulation model, including the proposed control architecture ("Improve" phase of LSS)
19. Implementation of the suggested optimization solutions on the simulation model ("Improve" Phase of LSS)
20. Calculating the system's timing as a result of the simulation model's optimization solutions ("Improve" Phase of LSS)
21. Determining the system's OEE percentage following modification ("Improve" phase of LSS)
22. Comparing the system's OEE percentages prior to and following modification ("Control" Phase of LSS)
23. Determining which resources' performance improved as a result of modification and the reason for the improvement. ("Control" Phase of LSS).

2 Literature Review

2.1 Overview

This section summarises previous research on the implementation of agent-based control architectures for SMEs with the goal of facilitating their adoption of Industry 4.0. Additionally, it provides a road map for successful enterprise evaluation and improvement by reviewing various types of literature regarding the most appropriate techniques, standards, and approaches for achieving the research objectives described in the previous section.

2.2 *Agent-Based Control Architecture for the Industry 4.0 Implementation*

Wan et al. (2016) demonstrated a smart factory that incorporates a variety of intelligent objects and products. According to the authors, integrating an enterprise with smart objects and products results in the implementation of a reconfigurable and flexible manufacturing system, both of which are characteristics of Industry 4.0. The authors of this study developed a multi-agent system that is compatible with cloud technology and capable of networking with intelligent shop floor resources such as machines, conveyors, and products [72].

Prinz et al. (2016) developed a methodology for connecting all resources associated with manufacturing processes. According to the authors, the main goal of Industry 4.0 implementation is to connect real physical resources. The authors also stated that new intelligent manufacturing control architectures should be used to achieve this goal. The authors proposed multi-agent control as the desired control system [73].

Santos et al. (2017) outlined a strategic roadmap for Industry 4.0 adoption, focusing on key technologies. The authors discussed eight topics, including cyber-physical systems (CPS), the Internet of Things (IoT), embedded systems, cloud computing, services-oriented manufacturing, sensing, agent-based control systems, additive manufacturing, and robotics [74].

Moeuf et al. (2018) conducted a literature review to analyze the production planning and control system of SMEs. The authors directed that the literature review adheres to the Industry 4.0 concept. The authors emphasized the importance of the agent-based control system due to its impact on the manufacturing system's flexibility. According to the authors, a critical factor in facilitating Industry 4.0 adoption is the manufacturing system's flexibility [75].

Bechtsis et al. (2017) presented an overview of the capabilities of material handling systems in Industry 4.0. The study examined the effect of a new control system for material handling systems, a critical component of SME operations. The authors conducted a thorough examination of the barriers that SMEs face in adopting Industry 4.0. The authors recommended that the most effective solutions include a proper control architecture and production planning [76].

2.3 *Agent-Based Control Architectures for Manufacturing System (Material Handling Systems Approach)*

Chan et al. (2011) proposed a model for improving material handling system called material handling equipment selection advisor (MHESA) [77]. Lewandowski et al. (2013) emphasized the importance of multi-agent communication between physical interfaces. The authors developed a methodology for assessing the benefits of multi-agent communication when applied to a material handling system within an enterprise [78]. According to the authors, they encountered a number of difficulties while implementing a new control architecture on a particular material handling system. External interfaces, legacy system integration, conflict resolution, overall system control, system dynamics, communication, system architecture, ontology management, and control architecture are the primary issues mentioned by the authors. They stated that the majority of the issues mentioned can be addressed by implementing a type of manufacturing control system that possesses the characteristics of an agent-based control architecture [77, 78].

Johnstone et al. (2010) applied the agent-based control concept. This concept is critical for avoiding material collisions in material handling systems. The authors recognized that an efficient layout of material handling lines was critical for the parts to travel in the correct direction without encountering any obstacles. Additionally, the gaps between the objects were maintained to achieve the desired differentiation at the merging point on the target MHS. The authors concluded that this can only be accomplished through the use of an agent-based control architecture to intelligently manage conveyor and input/output lines [79].

Lau and Woo (2008) developed an operator designed routing methodology for material handling system [80]. They proposed a material handling system consisting of agents connected in a single direction for the purpose of controlling the arrangement of MHS, which they dubbed collaborating node. They defined the most suitable route in terms of degree of tolerance for unexpected factors and material workload balancing in order to arrive at directing choices. The authors refer to this approach as distributed real-time state mapping and define directing procedures in general [80].

2.4 *Importance of Manufacturing System Performance Evaluation*

According to Jain et al. (2011), improved manufacturing performance results in increased enterprise competitiveness. The authors defined two stages necessary for a business to succeed in a competitive market. The first stage is to establish competitive priorities, and the second stage is to identify critical manufacturing factors that contribute to superior manufacturing performance [81].

According to Wakjira and Singh (2012), a proper measurement system for evaluating an enterprise's performance enables management to exercise comprehensive control over the enterprise and consequently improve production [82].

According to Hon (2005), the most critical subject for achieving an enterprise's objectives is the evaluation of the manufacturing system's performance. According to the author, there is a direct correlation between the enterprise control system and the manufacturing system's performance. Additionally, the author mentioned the possibility of extremely precise physical or mechanistic performance measurements, but he stated that enterprise performance evaluation may be an unsettled subject due to the multidimensional and diverse nature of manufacturing systems [83].

Abdi and Labib (2011) evaluated the efficacy of several conventional methods for evaluating the performance of manufacturing systems. According to the authors, it is critical to consider several manufacturing factors that are not considered in conventional evaluation methods. The authors demonstrated the critical nature of manufacturing system performance evaluations by examining their impact on reducing the enterprise's uncertainty caused by a variety of external factors [84].

Bol et al. (2016) investigated the relationship between the control system's design and managers' evaluation decisions and behaviour. According to the authors' findings, control system design elements have an effect on the accuracy of the information required to assess the transparency of performance evaluation [85].

2.5 *Lean Six Sigma Approach for Manufacturing System Performance Evaluation and Improvement*

Ramesh et al. (2016) demonstrated how Lean Six Sigma can be used to improve the Overall Equipment Effectiveness (OEE) of small and medium-sized enterprises. According to the authors of this study, SMEs should continuously increase and improve their performance efficiency and quality. Additionally, the authors stated that a low OEE resulted in high costs associated with product reworks. As a result, the authors proposed that six sigma implementations will identify and eliminate problems while also improving the manufacturing system's performance, thereby increasing OEE [14].

Albliwi et al. (2015) conducted research to determine how manufacturing enterprises can continuously improve their performance. According to the authors, Lean Six Sigma employs the most effective strategies for continuously improving the performance of manufacturing systems. The authors concluded that Lean Six Sigma approaches assist businesses in achieving operational and quality excellence and thereby improving performance [13].

Swarnakar and Vinodh (2016) used a business-oriented approach and a Lean Six Sigma strategy to improve the performance of manufacturing enterprises. The authors stated that by implementing Lean and Six Sigma-DMAIC strategies, defects will be reduced and non-value-adding tasks will be eliminated in the manufacturing system. Additionally, the authors stated that the Lean Six Sigma framework improves the target manufacturing system's key metrics [86].

Habidin et al. (2016) used a methodology to ascertain the relationship between the implementation of lean Six Sigma, the manufacturing control system, and manufacturing performance measurement. The author developed a mediator model to demonstrate the relationship between the presence of an effective manufacturing control system, Lean Six Sigma, and manufacturing system performance. According to the authors, when combined with Lean Six Sigma, a proper manufacturing control system improves manufacturing system performance [87].

2.6 *Time Study Techniques for Lean Six Sigma and Manufacturing System Performance Evaluation and Improvements*

According to Franchetti (2015), in order to analyze a manufacturing system, the production rate or system capacity, as well as associated cyclical patterns, must be determined. The authors established time study as a necessary component of data collection for analyzing the aforementioned subjects in order to analyze a manufacturing system. Additionally, the author stated that a time study demonstrates the manufacturing system's behaviour when completing some specific tasks. The author concluded that accurate time study data and information flow into a successful Lean Six Sigma evaluation. By conducting a comprehensive case study, the author demonstrated the importance of time study in the "Analyze" and "Measure" phases of Lean Six Sigma [88].

Lande et al. (2016) examined the critical success factors for small and medium-sized enterprises using the Lean Six Sigma framework. According to the author, identifying these critical success factors results in the development of a proper Lean Six Sigma framework that impacts the quality and performance of small and medium-sized enterprises. The authors intended to emphasize the critical nature of selecting an appropriate time study tool for identifying the most effective critical success factors necessary for effective Lean Six Sigma implementation [89].

2.6.1 Classic Time Study Techniques

Al-Saleh (2011) used process charts, flow charts, and activity charts to detect work processes involving handling, operation, storage, and delays [90]. Gruzauska et al. (2016) examined flow charts that were also used to indicate transportation, preparation, and shift times [91].

Gilad (2006) conducted a time study in a textile factory to determine the standard time required to operate a printing machine in order to calculate the performance rating using the speed-rating technique. The author divided three operations into small elements using a methodology that included a classical observation. The author concluded that the most time-consuming operations are those that require operators to maintain constant attention. Additionally, the author stated that variables were not controlled during the time study observation, and the study focused exclusively on machine and related operations, as well as human labour and its performance. As a result, the author concluded that the study cannot be used to benchmark another system with distinct variables [92].

Gruauskas et al. (2016) conducted a study with the objective of analyzing the manufacturing process through the use of labour and machine productivity and flow process charts. According to the authors, there is a great deal of idle time as a result of some lengthy processes that run in parallel with other activities. As a result, they have proposed that certain independent activities be carried out concurrently and in parallel with one another [91].

Bon and Daim (2010) examined time as a metric for evaluating an enterprise's performance. Time is a unit of measurement that is used to organize the tasks available in an enterprise. The purpose of this study was to increase production while decreasing its cost, which involves a large number of time-consuming operations [93].

2.6.2 Method Time Measurement as a Predetermined Time Study Technique

Longo and Mirabella (2009) used a different tool to design a manufacturing assembly line. Due to the study's reliance on an illusory assembly line, a simulation tool was used. To simulate the workplace with a qualified worker, another simulation tool was used in conjunction with Method Time Measurement (MTM) as a time study technique to analyze the assembly line's primary model. The findings indicate that some stations have bottlenecks, others have an inefficient distribution of tasks, and still others have issues with labour ergonomics and working conditions. Several modifications were made to establish the optimal work method, line balancing, and risk-free ergonomics [94].

Kuhlang et al. (2011) presented a study on an assembly line and production-logistic process with the goal of generating additional added value elements during the process within a specified time frame. The authors are aiming for a faster process with increased production time rather than for increased production in a fixed/constant duration. To accomplish this, a combined methodology is used, which includes MTM as a hybrid optimization of added value. Due to the methodology's detailed analysis, the results indicated a high rate of time identification and elimination. As a result, a more efficient process with a shorter lead time and increased productivity can be designed [95].

De Almeida and Ferreira (2009) used MTM to conduct mythology on two automotive and household appliance manufacturing companies. MTM is primarily used to create timetable data that aids in planning, organizing, and utilizing available resources. According to the authors, MTM is an important tool for identifying unnoticed small-time wastes. Additionally, they stated that MTM is not a competitor to nor an alternative to other time study tools, but rather a complementary tool that generates only timetables [96].

2.6.3 Maynard Operation Sequence Technique Time Study

Karim et al. (2016) examined the efficacy of using the Maynard Operation Sequence Technique (MOST) as a time study technique in an automated enterprise. MOST captured video of available stations and examined tasks to make system improvements. By implementing MOST and standardizing the time required to complete each task and the entire system, the cycle time was significantly reduced [97].

Gupta and Chandrawat (2012) examined the performance of a Japanese SME that relies on mass production to enable businesses to produce more products with fewer resources. According to the authors, MOST is the more practical method because it can identify non-added value work through measurement and can be applied to any type of task [98].

2.7 Overall Equipment Effectiveness for Lean Six Sigma and Manufacturing System Performance Evaluation and Improvements

Gibbons (2006) developed a methodology that incorporates Overall Equipment Effectiveness (OEE) as a measure of enterprise effectiveness in order to align it with the Lean Six Sigma-DMAIC improvement methodology. The author demonstrated the value of OEE by providing data on enterprise potentials. The author asserts that for the Lean Six Sigma process, failure mode and effect analysis may be unnecessary. The author concluded that OEE can be used to analyze the system's captured data, which is necessary for performing an effect analysis and determining the system's potential [99].

Gibbons (2010) introduced a framework for assessing the capability of Six Sigma processes using OEE data. According to the author, optimizing OEE combines asset management effectiveness, net process performance, gross process performance, process effectiveness measurements, and Six Sigma process capability into a single lean process. Six Sigma is a critical performance indicator for assessing an enterprise's performance [99].

Mandahawi et al. (2012) published a study aimed at enhancing the performance of manufacturing firms through the use of Lean Six Sigma strategies. The authors improved the target manufacturing system's productivity by implementing the Lean Six Sigma-DMAIC methodology and a variety of lean tools. Two performance indicators, namely Overall Equipment Effectiveness and production rate, were used to assess the manufacturing system's performance prior to and following the Lean Six Sigma-DMAIC cycle [100].

2.8 Overall Equipment Effectiveness and Time Study Relationship

Ghafoorpoor Yazdi et al. (2018) developed a methodology for evaluating the Overall Equipment Effectiveness (OEE) of a material handling system by using the results of a comprehensive time study. According to the author, the OEE tool is intended to identify equipment losses that reduce equipment effectiveness. The author concluded that these losses are caused by tasks that consume resources but do not generate value. According to the authors, there are six major losses in manufacturing systems: distance between resources due to layout design constraints, downtime, speed, idling, minor stoppage, and reduced speed losses [101].

Puvanasvaran et al. (2013) used a time study method to conduct a study aimed at increasing the overall equipment effectiveness of autoclave machines in the aerospace industry. The study was conducted in two stages. The first step is to calculate the current OEE percentage using a stopwatch time study. The second step is to increase the OEE percentage by utilizing Maynard's Operation Sequencing Technique (MOST). The authors assert that the MOST are critical for identifying value added and non-value added activities at each stage of the process [102].

Patel (2015) conducted a time study using a stopwatch. The authors stated that using a stopwatch is always the best method for examining manual labour because human performance is not always consistent. The author determined the cycle time for manpower tasks and, using the overall equipment effectiveness (OEE) standard, reduced the overall shift's cycle time. After conducting a time study, standard times were established, and through appropriate actions, cycle times in operations were reduced, as well as the number of human labourers [103].

2.9 Computer-Based Simulation for Lean Six Sigma and Manufacturing System Performance Evaluation and Improvements

Yang et al. (2003) covered all critical organizational, implementation, technical, and philosophical aspects of developing a proper Six Sigma. The authors established computer-based modelling and simulation as a viable concept for developing effective Six Sigma strategies. According to the authors, computer-based simulations aid in ensuring the efficacy of new optimization concepts that are subjected to Six Sigma's "improve" phase. Additionally, the authors stated that utilizing computer-based simulation enables them to ensure that new process optimization concepts generate the appropriate functional requirements [104].

Naeem et al. (2016) used a simulation model to estimate the amount of storage space required in the fabric manufacturing industry. In this research, the author used simulation software to create a model that accurately represents a real system, complete with all physical components and available tasks. The authors applied their methodology to the simulation model in order to improve the performance of the real system [105].

Jayant et al. (2012) used Arena software to create a simulation model to predict the system's future performance. The authors determined resource utilization, cycle time, transfer cost, and transfer time in this study [106]. Zahraee et al. (2014) used Arena simulation software to investigate the development of company productivity. The optimal factors that have a significant impact on the production of the company were identified in this study through an analysis of the Arena simulation outcomes [107].

Maropoulos and Ceglarek (2010) illustrated the product lifecycle concept of verification and validation. The authors validated and verified their techniques by analyzing the simulation model's graphical behaviour and comparing it to the real system. The authors stated that the study was expanded to include every component of the model through the analysis of output data, which resulted in effective and efficient results [108]. In a study using supported techniques, Sargent (2013) summarised the verification and validation process succinctly. Additionally, the author suggested that visualizing the model's operational behaviour as it moves through time as a useful technique for validation [109]. Macal (2016) provided a detailed explanation of the validation and verification of an agent-based system using a simulation model and an analysis of the simplest realistic rules governing its behaviour [110].

2.10 Literature Review Conclusion

The section's first objective was to review the road map for Industry 4.0 adoption by Small and Medium-sized Enterprises and to establish the principles for implementing Industry 4.0. Numerous studies conducted over the last decade on the principles of Industry 4.0 were reviewed.

According to the studies, small and medium-sized enterprises (SMEs) require a new collaborative, agile, and responsive control architecture to enable industry 4.0 adoption. As a result, the review of the literature was limited to agent-based control architectures applicable to SMEs, with Master–Slave being the most appropriate agent-based control architecture. Similarly, demonstrated the constraints, difficulties, and uncertainties encountered by SMEs when modifying their current control architecture. Numerous researchers have attempted to define Lean Six Sigma-DMAIC strategies for evaluating and improving the performance of small and medium-sized enterprises (SMEs). As a result, Lean Six Sigma was chosen and the best techniques for the associated phases were evaluated (Define, Measure, Analyze, Improve and Control).

It was concluded that an appropriate data collection technique is required to provide the required system data for the Lean Six Sigma "Measure" phase. As a result, an appropriate combined time study technique was chosen. Additionally, Overall Equipment Effectiveness (OEE) was chosen for the Lean Six Sigma "Analyze" and "Improve" phases. The "Measure" and "Analyze" phases of Lean Six Sigma assist in identifying issues, constraints, and potential solutions for improving system performance following the implementation of the proposed control architecture. All of these potential solutions should be investigated in order to resolve the system's identified issues and limitations. Discrete event simulation, according to the literature, is the appropriate tool for this purpose. As a result, a simulated version of the real system was used to implement these solutions and modifications. The outcome of this simulation was used in the Lean Six Sigma "Measure" and "Improve" phases.

Following a review of the most recent literature on each of the research objectives, it was determined that there is a disconnect between the requirements of SMEs and Industry 4.0 principles. This gap exists because of the characteristics of SMEs and their inability to modify their current manufacturing systems. The proposed novel agent-based control architecture, as well as the methodology for evaluating and improving the performance of SMEs, enables the adoption of Industry 4.0 to close the aforementioned gap.

3 Methodology

3.1 Overview

This section used a case study methodology to demonstrate the process of designing and implementing a novel agent-based control architecture for SMEs. The purpose of this study was to develop a novel roadmap for evaluating and improving the target system's performance following the implementation of the proposed novel control architecture. Evaluation and improvement techniques are used in conjunction with Lean Six Sigma strategies. Time study, overall equipment effectiveness (OEE), and discrete event simulation are used to gather system data and information for the various Lean Six Sigma phases.

3.2 Design of the Agent-Based Control Architecture with Master-Slave Mechanism

The agent employs a variety of computing paradigms, and each of these paradigms has a number of theoretical and technological foundations. These paradigms and their associated requirements can be classified into two broad abstractions:

- An agent is a computational system capable of autonomous and intelligent behaviour.
- An agent is capable of interacting with other agents.

To adhere to the abstractions for an agent mentioned in the proposed novel control architecture, an agent-based control architecture with a Master–Slave mechanism was chosen as the target in this research.

3.2.1 Structural Definition of the Proposed Control System

The proposed agent-based control architecture was designed with task division, allocation, and communication patterns as primary considerations. To delegate tasks to the agents, a master–slave mechanism was required. The proposed agent-based control architecture with Master–Slave mechanism was composed of three distinct layers: the physical resource layer, the physical resource control layer, and the management layer. Required features were defined for each layer based on the characteristics of an agent-based control architecture with a Master–Slave mechanism. Additionally, communication and interaction protocols between layers were defined with the required network and its associated interfaces in mind.

Physical Resource Layer of the Proposed Control Architecture

This layer encompasses the category of physical resources, which encompasses all operational resources concerned with the physical capabilities necessary to perform the required tasks. Due to the target system's primary operational task of material handling in accordance with the planned scenario, the majority of resources are material handling equipment such as robots, conveyors, and sliding units. Apart from the physical resources required for material handling, another category of physical resources was required to support the detection and measurement tasks. Different types of sensors were used to detect the presence of material in specific locations on material handling systems. Along with separating the materials on a material handling system, certain sensors were used to determine the required physical variable for detecting the object's colour (Fig. 3).

Resource Control Layer of the Proposed Control Architecture

As illustrated in Fig. 3, this layer of the proposed control architecture has two levels or layers of control. At the lower layer, there is a set of control units known as Slave agents that are responsible for the control of the physical resource layer's resources; and at the top layer, the Master agents control unit is responsible for the coordination of slave interactions and communications.

Slave Agents in the Resource Control Layer

Slave agents were primarily responsible for controlling the associated resource/s within the target system. These agents are capable of performing a variety of tasks

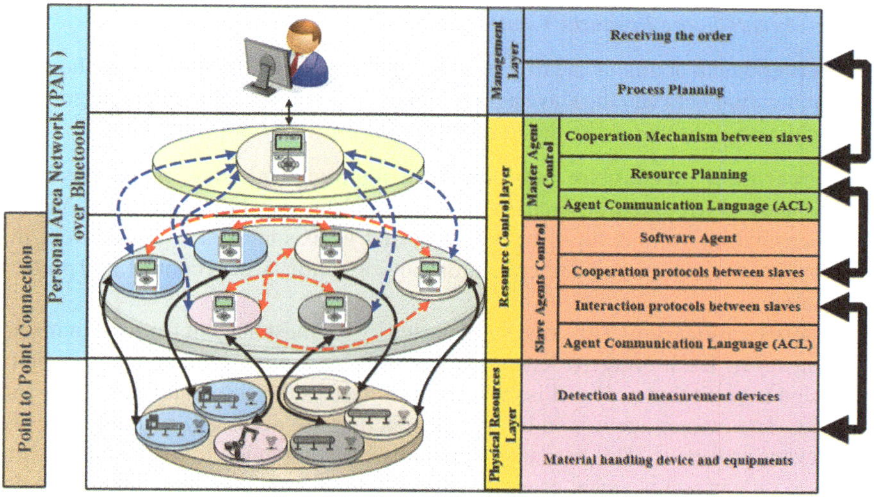

Fig. 3 Proposed agent-based control architecture with master–slave mechanism

that must be defined, created, and assigned to them prior to being used in the system. The control architecture was designed to invoke the task in response to the system's requirement to execute an order and to facilitate collaboration with other slaves via communication and cooperation protocols. As a result, a slave agent consisted of the following:

- **Software agent:** In this research, a software agent is defined as an intelligent control software that consists of a programme for controlling the associated resource/s and a programme for managing cooperative interactions with other slaves and master agents. This interaction is occasionally competitive with the other slaves.
- **Interaction protocols:** Each slave agent in this research adheres to a unique protocol/protocols in order to interact appropriately with other slaves and the master agent. For example, responding to messages from other slave agents, carrying out tasks assigned to them, or transferring their local states to other slaves and the master. Thus, these protocols were viewed as a means of defining the policies that slave agents should adhere to when interacting with one another and with the master agent.
- **Agent Communication Language (ACL):** Communication between Slave and Master agents was critical for realizing the agent-based paradigm's potential in the proposed control architecture. Slave agents required an Agent Communication Language to communicate with one another and with the master agent in order to exchange information and knowledge. The Agent Communication Language used in this research was the Foundation for Intelligent Physical Agents (FIPA).

Master Agents in the Resource Control Layer

The master agent, acting as the dispatch for all slave agents, facilitates collaborative communication between the slave agents and establishes the ideal task configuration for each slave. Consequently, the master agent comprised the following:

- **Cooperation mechanism between slave agents:** Each slave agent possesses the protocols for sending and receiving messages as well as the functions for encapsulating various types of data required to execute tasks. In addition, the master agent implements authentication, authorization, access control, and privacy features.
- **Resource allocation:** The order is requested by the master agent from the management layer. Each order for each product triggers a process of dynamic system design to fulfil that order. This procedure comprises the associated resources and tasks. The master agent is in charge of selecting the necessary resources for the relevant processes. This research's goal system is intended to differentiate parts based on a preset physical property (colour). There are four recognized steps for achieving part differentiation, with each phase requiring a distinct set of resources and activities.

Management Layer of the Proposed Control Architecture

In this research, the management layer receives the quantity of parts with a certain attribute (colour) and the related buffers, which are the system's exit locations for unloading these parts after differentiation. In this layer, the order will be assessed and the necessary processes to fulfil the order will be defined. In consideration of all essential procedures for order execution, a process plan will be produced and transmitted to the master agent (Figs. 4 and 5).

Communication Between Control Architecture Layers

Specific networks and connection protocols facilitate communication between the physical resources, agents, and management layers in the proposed control architecture. A peer-to-peer link was suggested to fulfil the communication need between slave agents and the physical resource layer, which consists of material handling devices and measurement or detection equipment. A Personal Area Network (PAN) was established to permit communication between slave agents in the physical resource layer and to provide communication protocols for them by the master agent. In this research, slaves and master communicate via Bluetooth connection across a PAN (Fig. 6).

3.3 Case Study System Description with the Agent-Based Control Architecture

To implement the suggested agent-based control architecture with a master–slave mechanism, a manufacturing system with the characteristics and features of small and medium-sized enterprises (SMEs) was required. Therefore, a modest instructional manufacturing system housed at the German University of Technology in Oman was chosen. The system consisted of several Material Handling System components, including conveyors, robot arms, and sliding units. The target system consisted of a small number of separate controllers that lacked specialized communication capabilities. The sole responsibility of the system was to handle the loaded material based on a predetermined sequence established by the operator.

3.4 System Description with the Agent-Based Control Architecture

To implement the suggested agent-based control architecture on the intended material handling system, certain minor hardware modifications were required, such as the addition of more sensors. In addition, a thorough task specification that included the required resources to complete the target tasks was necessary.

Figure 5 depicts the system layout design after the relevant sensors have been integrated. The system comprises a main conveyor with four sensors, a robot arm, two side-conveyors (left and right) with motion sensors and two sliders (left and right) with colour sensors. Two sets of servo motors (Four Motors) operated the conveyors, while each slider was powered by one motor. Individual slave agents managed the conveyors, sliders, and robot. The slaves received input signals from sensors in order to generate output signals for the resources. Figure 4 depicts both the master and slave agents.

To efficiently perform communication, cooperation, and the transfer of knowledge between slaves and master agents, all agents were communicating simultaneously through a shared perception network (Fig. 6). The Master agent task in the system required a protocol to guide and support the interaction between slave agents and the master agent, which was specified as the Master agent task.

Using Fig. 5 as a guide, the system can be divided into four major districts or phases. The initial phase includes four sensors, material storage (Red, Blue, Yellow, and Green), a primary conveyor, and a slave agent1. Phase 2 includes the robot arm and slave agent number 6. Phases 3 and 4 are similar, with the exception that they are allocated to handle distinct items in the opposite direction. Both phases 3 and 4 consist of two slider units equipped with a colour sensor for each and controlled by slaves (Slave agent 4 for the right and Slave agent 5 for the left), two conveyors equipped with a motion sensor for each and controlled by slaves (slave agent 2 for the right and Slave agent 3 for the left), and unloading buffers. The unloading buffers

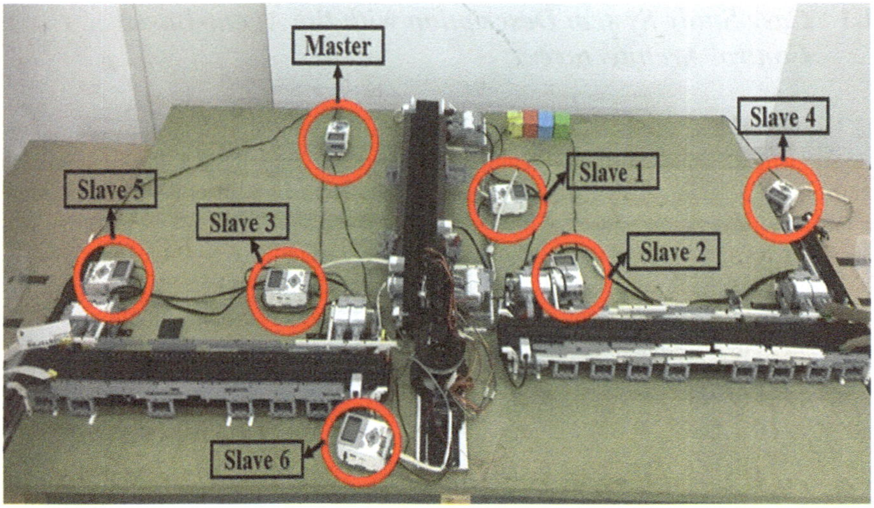

Fig. 4 System overview

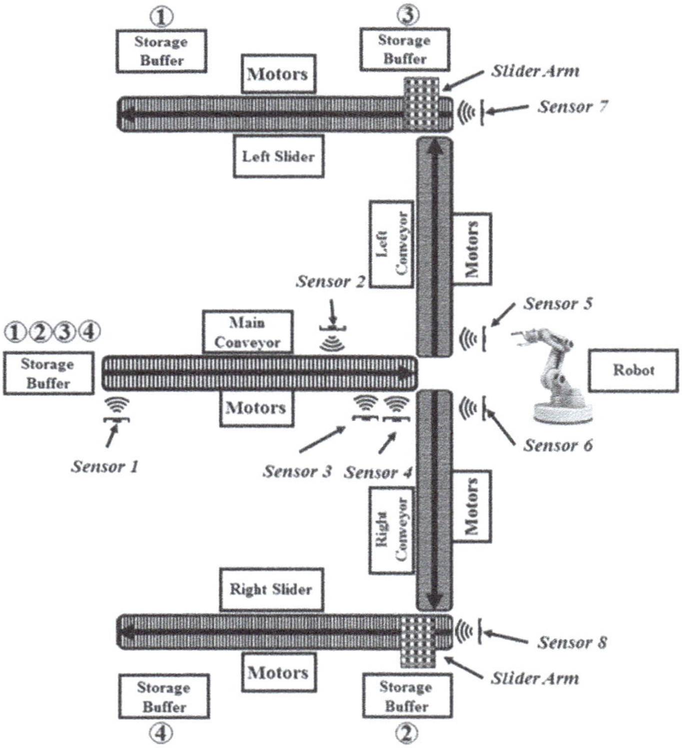

Fig. 5 Layout design—top view

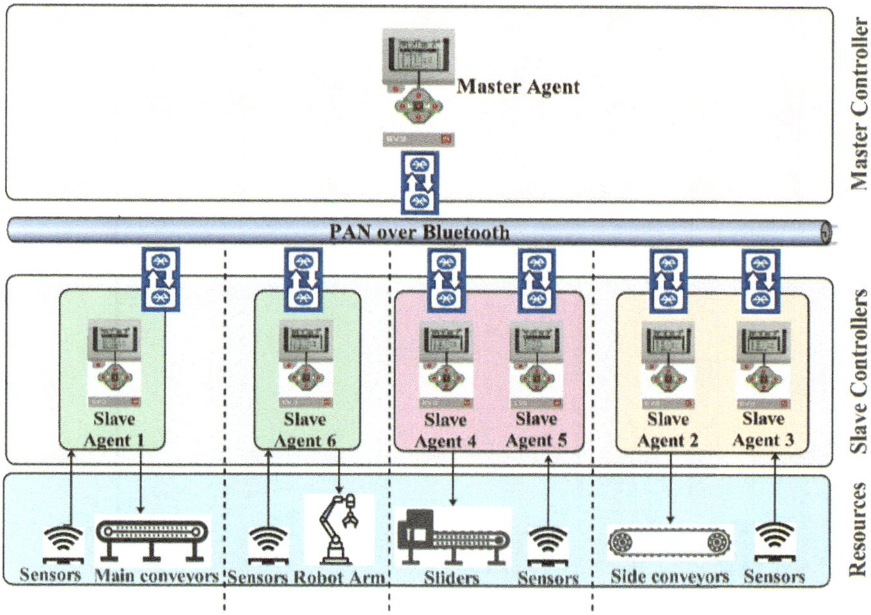

Fig. 6 Communication in the proposed agent-based control architecture

are the outlets/exits of the target system, and materials will be unloaded separately in each one based on their colour and the scenario that has been designed.

Each phase of the target system possesses unique functionalities and control architectures. In the initial phase, the primary conveyor was initiated by sensor 1 (Infrared sensor). Sensor 2 (an infrared sensor) was added to the system in order to slow down the conveyor in order to precisely position the objects. By slowing the conveyor, sensor 3 was given sufficient time to detect the existence of the object, hence increasing the system's precision. The robot arm was designed to remove the material from a predetermined location on the main conveyor and place it on the left or rightside conveyors (Fig. 7). Therefore, it was required to reduce the speed of the primary conveyor motion in order to increase the precision of sensor 2 and, ultimately, sensor 3, in order to stop the conveyor and object in the exact desired location (Fig. 5).

A robot arm is the primary resource for the second phase. The robot arm's task is to remove the object off the main conveyor and place it on the right- or left-conveyor based on the object's colour and the plan for that colour (Fig. 8).

As the objective of the suggested agent-based control architecture was to provide distributed control over each resource, two separate controllers were assigned to the main conveyor and robot arm, respectively. In order to improve the intelligence agent software, additional sensors were put into the system for each slave agent. Taking into account the most essential aspect of agent-based control architecture, a

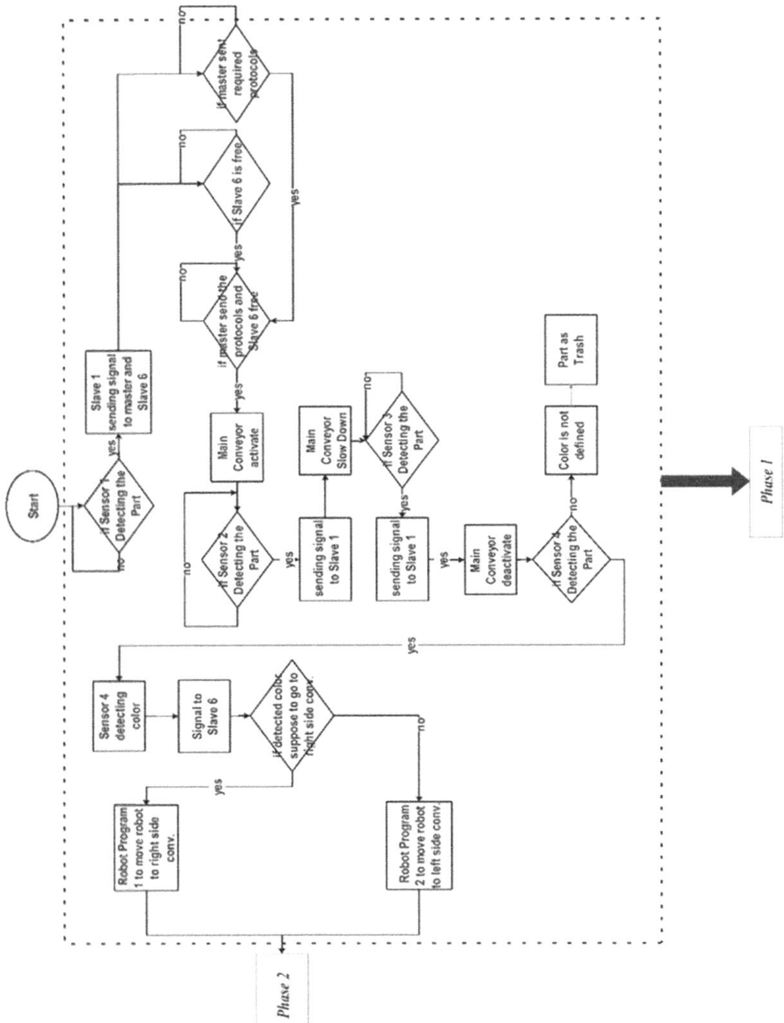

Fig. 7 Main conveyor and robot arm control algorithm (Phase 1)

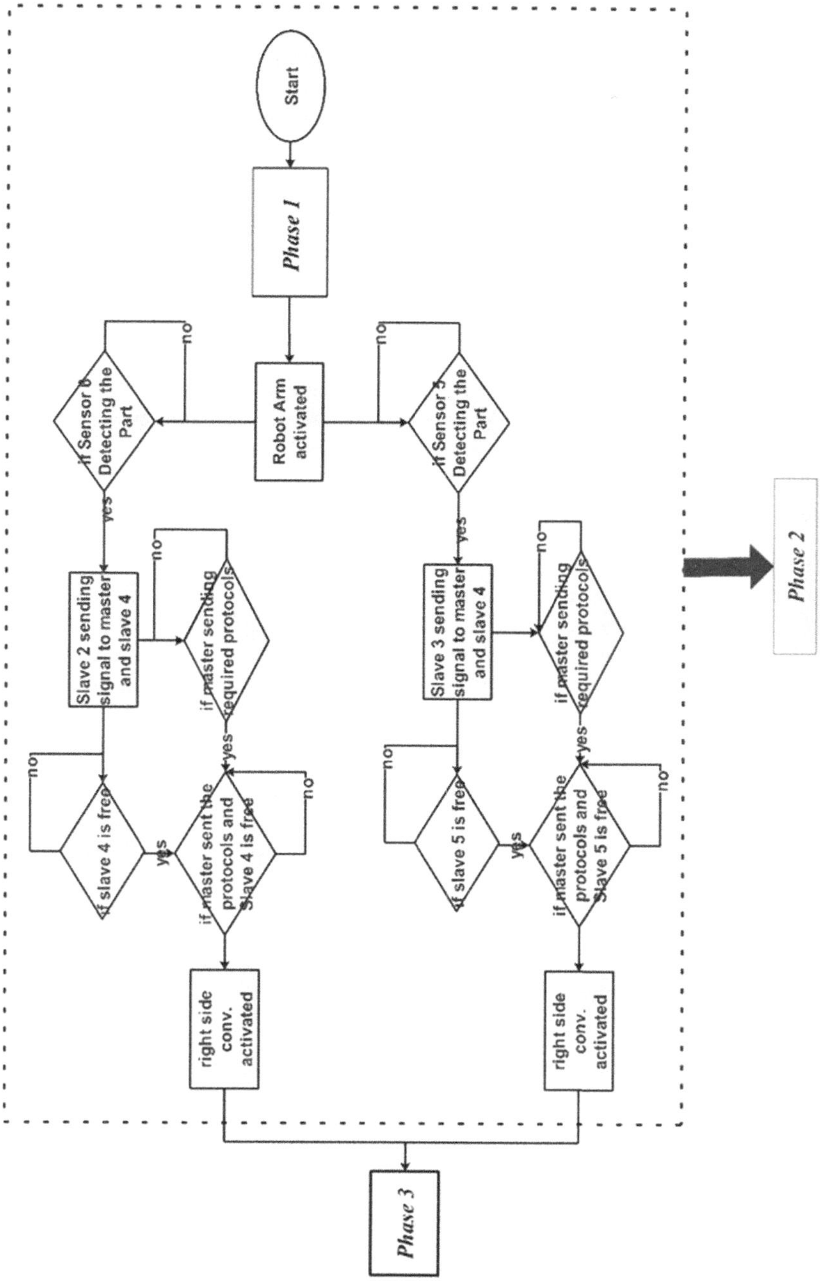

Fig. 8 Robotic arm and side conveyors control algorithm (Phase 2)

communication algorithm was designed between slave agent 1, slave agent 6, and master agent in order to improve the interaction between the main conveyor and the robot. To implement the described communication algorithm between Phase 1 and Phase 2 (Main conveyor and robot arm), the necessary communication and interaction protocols on Master agent and associated slave agents were defined (Figs. 9 and 10).

In phases 3 and 4, two sensors were added to the left and right conveyors' beginnings. These sensors were used to trigger the side conveyors (Fig. 5). Since the object must slip to the slider for colour-based differentiation, once the object slides to the slider, the sensor on the slider differentiated the colour of the detected object and allocated the object to the target buffer accordingly (Fig. 9).

Similar to Phase 1 and Phase 2, a communication algorithm was necessary between Phase 2–Phase 3 and Phase 2–Phase 4 (Fig. 11). These algorithms enhance cooperation amongst the following groups of agents:

- Master agent, slave agent 6 (robot arm) and slave agent 2 (right side-conveyor).
- Master agent, slave agent 6 (robot arm) and slave agent 3 (left side-conveyor).
- Master agent, slave agent 2 (right side-conveyor) and slave agent 4 (right sider unit).
- Master agent, slave agent 3 (left side-conveyor) and slave agent 5 (left sider unit).

The suggested agent-based algorithm in conjunction with the communication algorithm enables the system to regulate the functioning of the resources based on a predetermined scenario, hence preventing task overlap and conflict. For example, slave agent 1 associated with the main conveyor activates the conveyor and communicates with slave agent 6 linked with the robot arm to reactivate it and transfer the next part. While slave agent 6 is active and performing its assigned task, slave agent 1 will not receive the confirmation signal until the task is complete. The master agent will supply and control all of the aforementioned communication and interaction protocols between the slave agents.

3.5 Lean Six Sigma Strategies to Evaluate and Improve the Performance of the Target System After Agent-Based Control Architecture

As the main part of the objective for this research, the performance of the target manufacturing system (material handling system) after implementation of the proposed agent-based control architecture is evaluated. Furthermore, after evaluation of the performance of the system, limitations and problems of the system were identified. Therefore, for each of the identified problems and limitations, a proper solution was proposed to overcome their negative influences on the performance of the system and improve it accordingly.

In order to reach this goal, Lean Six Sigma-DMAIC strategies were selected as an accurate method to identify and improve the performance of the system. According to

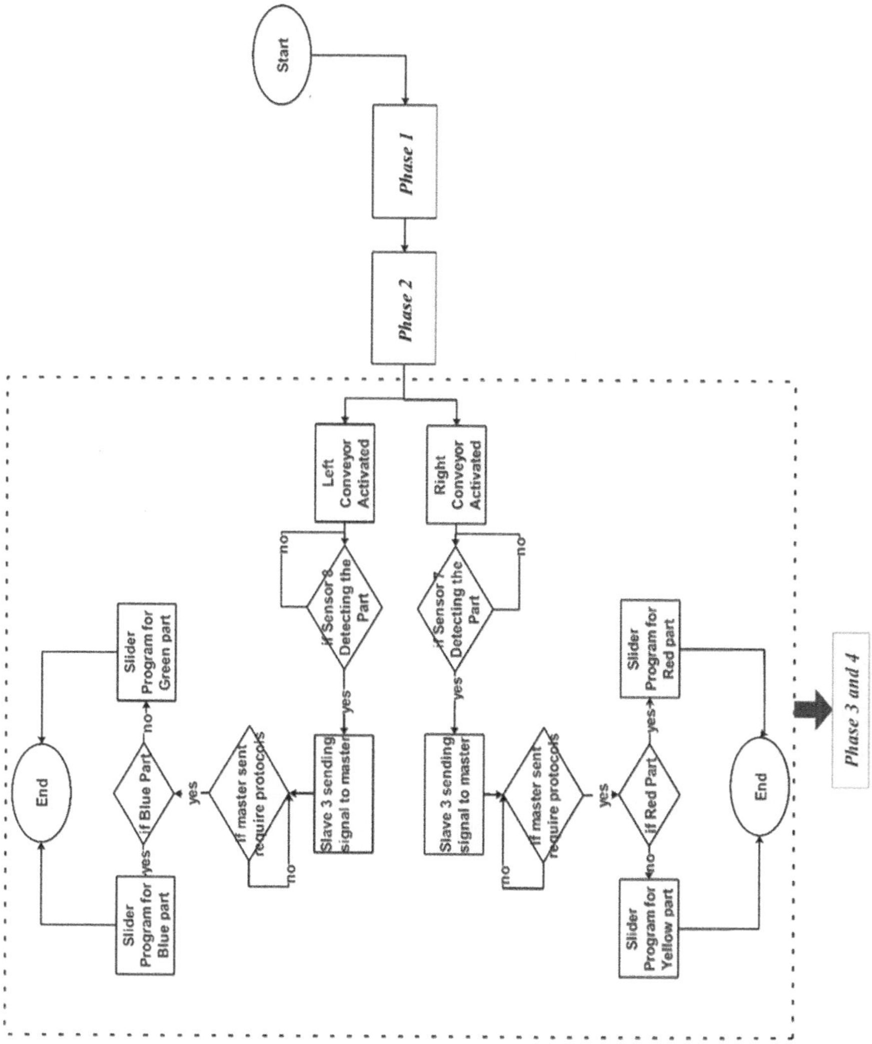

Fig. 9 Side conveyors and sliders control algorithm (phases 3 and 4)

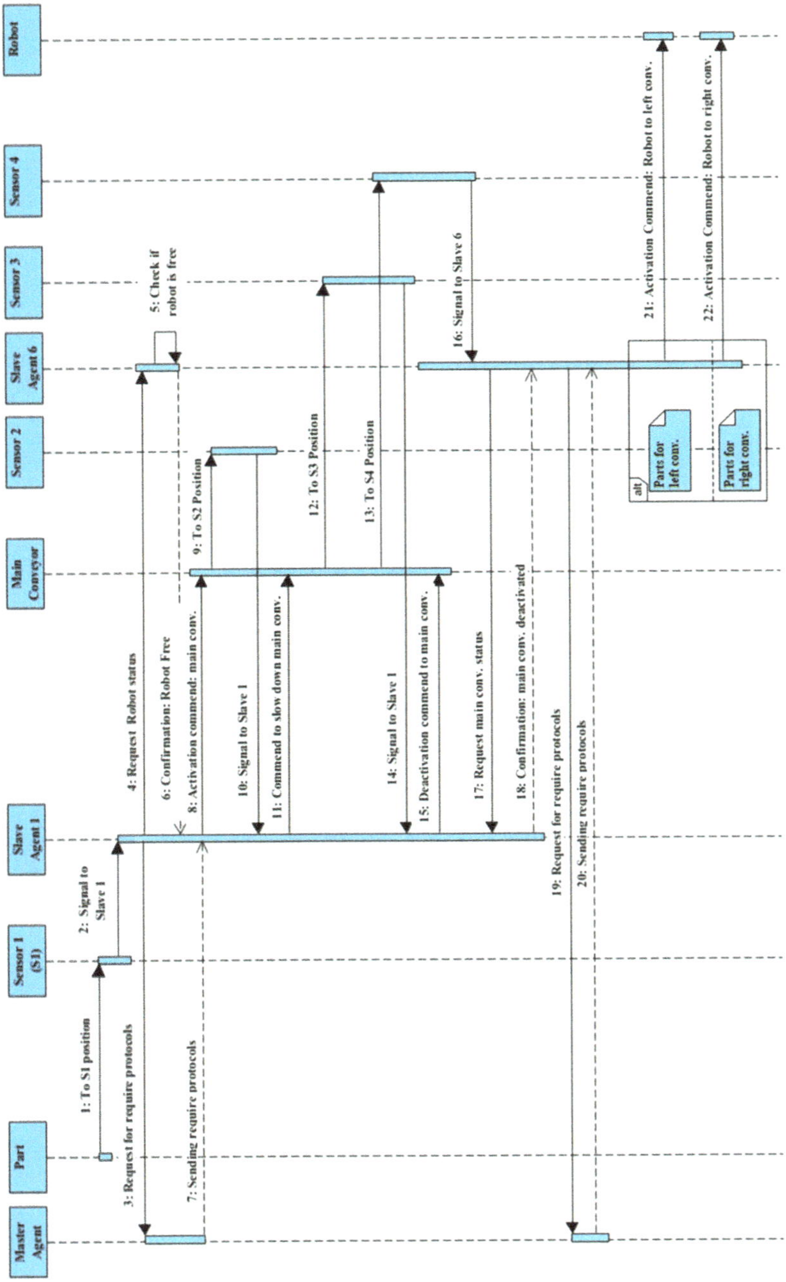

Fig. 10 Communication algorithm between slave agent 1 (main conveyor) and slave agent 6 (robot arm) and Master agent

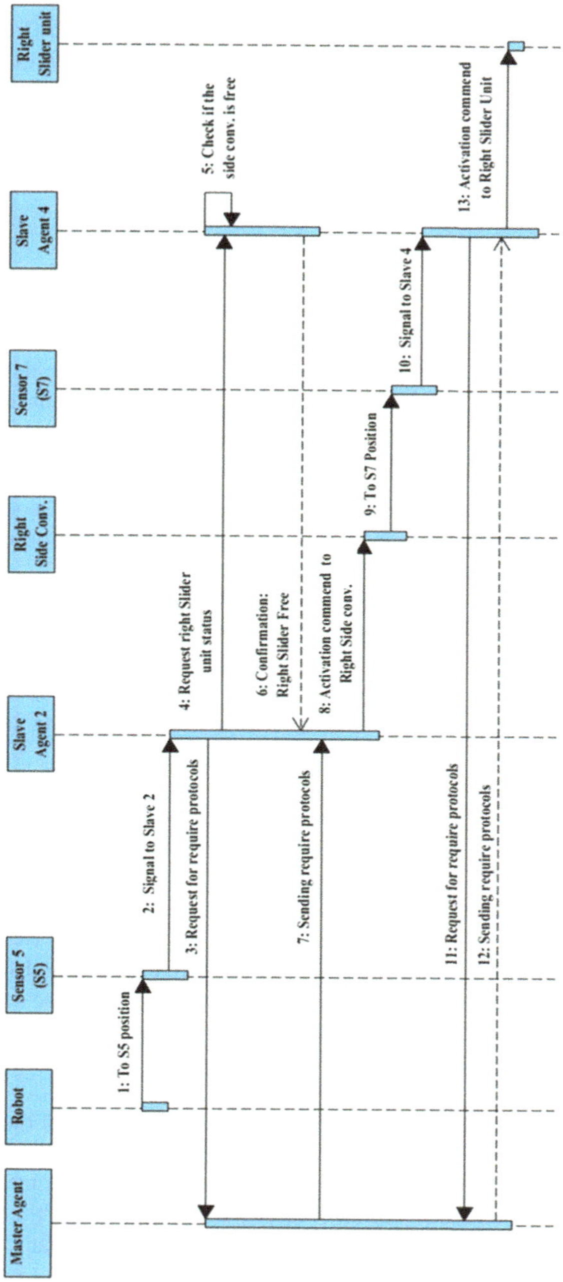

Fig. 11 Communication algorithm between slave agent 6 (robot arm), slave agent 2 (right side-conveyor), slave agent 4 (right slider unit) and master agent

the definition of Lean Six Sigma, there are five individual phases which are "Define", "Measure", Analyze", "Improve" and control. To follow the Lean Six Sigma strategies, the required information, data, methods, and tools were defined. This procedure required a comprehensive consideration of the target system's properties and characteristics. As the system has been considered as SMEs, the requirement for each phase of Lean Six Sigma-DMAIC were provided by utilizing the most suitable methods, techniques and standards to obtain the required Information and data (Fig. 12).

3.5.1 "Define" Phase of Lean Six Sigma

In order to adhere to the strategies of the "Define" phase of Lean Six Sigma, certain crucial features of the target system were taken into account. The initial item under consideration was a complete problem statement for the desired manufacturing system (material handling system). This problem statement must address the following:

- A description of the problem(s) that the system may encounter during the process. Hardware (manufacturing resources), Software (agent software), and layout design are the primary categories of challenges and limits in the target system.
- The next step necessary for this phase is the definition of the manufacturing system's resources. Because the target system in this research is an example of a small and medium-sized enterprise (SME), there are limited resources such as material handling equipment (conveyors, robots, slider units), measurement equipment (Sensors), and control units (Master and Slaves including their software agents) are available.
- In this phase, the capability, functionality, and tasks of each resource must be defined. Material handling equipment in the target system is capable of transporting the material to different sections of the system and classifying the material based on the desired plan and control logic. The task of measurement equipment is to measure changes in physical variables and provide this information to control units as required input. The functionality of each resource is directed by control units, which provide timing and control signals based on the stated scenario.
- The manufacturing system's process plan should provide a concise description of how, when, and in what order the resources' tasks are to be done. Additionally, identify the appropriate channels of communication and interactions amongst the resources.
- During the "define" phase of Lean Six Sigma, the layout design of the production system must also be taken into account as an essential component. The layout is defined by providing the coordinates of each resource, its orientation as either horizontal or vertical, and the location of its crucial points, such as the load/unload point. Figure 5 illustrates the layout design of the intended material handling system for this research.

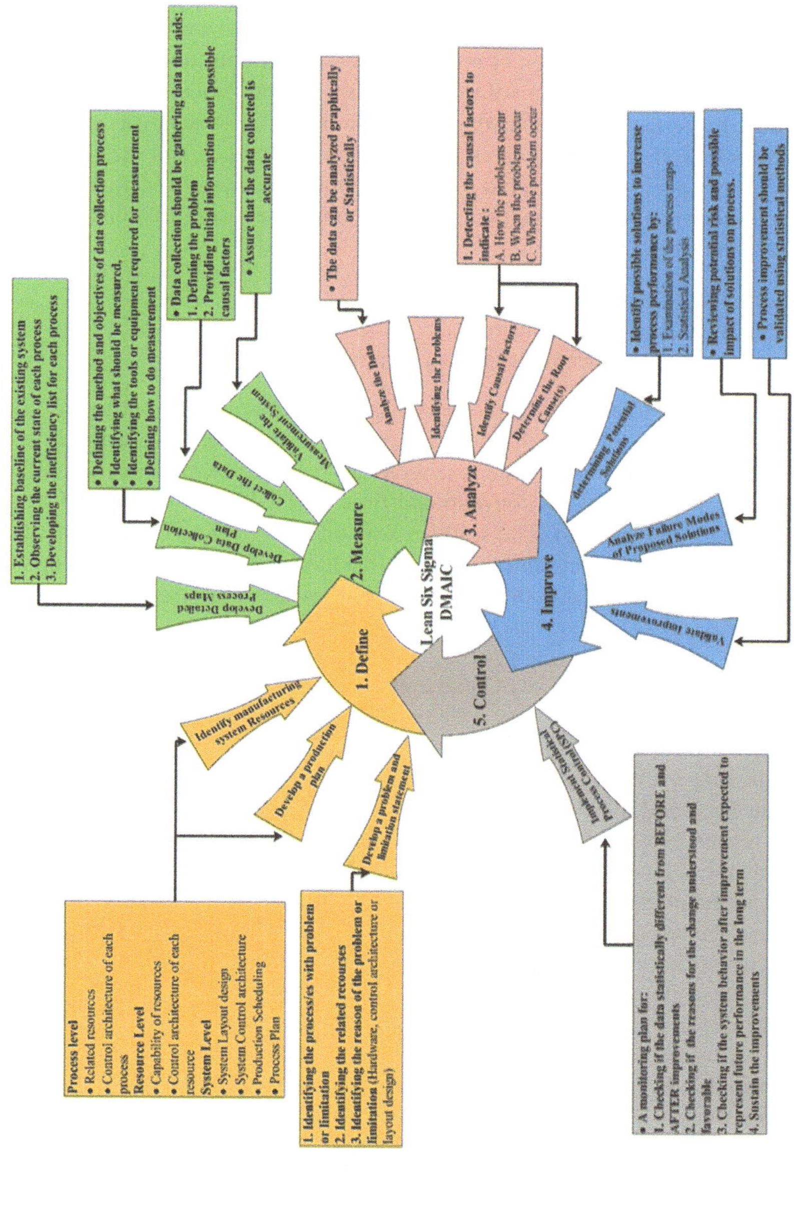

Fig. 12 Schematic procedure of lean six sigma implementation

3.5.2 "Measure" Phase of Lean Six Sigma

To meet the criteria of the "Measure" phase of Lean Six Sigma, a detailed plan for data gathering must be developed. This plan should include the important phases of the manufacturing process, as well as the corresponding tasks and resources. This strategy assists in the identification of process issues, constraints, and inefficiencies, such as extended cycle times, non-value added tasks, and bottlenecks.

To achieve this research objective, the target manufacturing system was separated into distinct sections. Each part contains a list of specific resources, and each resource provides a list of specific tasks (Table 1). After defining the system sections, tasks, and resources for each section, the data gathering objectives were established. The following are the goals:

- Identifying what has to be measured
- Identifying what has to be measured
- Selection of accurate procedures and approaches.

"Busy time" and "Idle time" of the resources were chosen as the criteria to be measured in this study. During the performance of the defined task/s, these two criteria were evaluated for each system section and resource. A combination Time study technique comprising the conventional time study, MOST, and MTM time study techniques were adopted. The software known as "ProTime Estimation" has been utilized as the most suited time study tool that has been capable of covering all required aspects of mentioned both time study techniques.

Time Study Methodology and Data Collection

Using "ProTime Estimation" software, the system was exposed to a time study test to determine the time cycle for each process component. The goal of the time study was to acquire the Busy and Idle time for each resource and undertake a thorough comparison of the resources with the same functionality. The system's time study assisted in identifying the resources or tasks with limitations and problems related with hardware (material handling equipment), software (agent software), and system layout design. These accomplishments as a consequence of time study assisted in focusing on specific resources with more problems and restrictions and finding appropriate optimization strategies and methods to increase system performance.

Figure 13 depicts the technique for collecting time study data using "ProTime Estimation" software. Three video files have been recorded from distinct system observation angles (Right, Left and Front view). The videos were uploaded to the software for time study. The time study analysis was conducted following the described methodology, where each recorded video was uploaded into the software, and for each video, the described processes for making time study reports were implemented (Fig. 14). Each of the possible tasks is marked with a number (Task Number) and the required resources to complete that task. These tasks were defined in the software

Table 1 Sections, task description and related resources for the entire scenario

Task	Task description	Included resource					
		Main conv.	Robot	Right conv.	Left conv.	Right slider	Left slider
1	Main conveyor handling red object to robot	✔					
2	Robot arm picking the red object from main conveyor		✔				
3	Robot arm placing red object to right conveyor		✔				
4	Right conveyor handling red object to right slider			✔			
5	Robot arm moves to its home position after placing red object		✔				
6	Main conveyor handling blue object to robot					✔	
7	Right slider transfers red object to red buffer					✔	
8	Right slider unloading the red object to red buffer					✔	

(continued)

Table 1 (continued)

Task	Task description	Included resource					
		Main conv.	Robot	Right conv.	Left conv.	Right slider	Left slider
9	Robot arm picking the blue object from main conveyor	✔					
10	Robot arm placing blue object to left conveyor		✔				
11	Right slider moves to its home position after unloading red object		✔				
12	Left conveyor handling blue object to left slider				✔		
13	Robot arm moves to its home position after placing Blue object		✔				
14	Main conveyor handling yellow object to robot						✔
15	Left slider transfers blue object to blue buffer						✔

(continued)

ProTime Estimation. After defining the tasks, the order of the tasks (Predecessors) was specified as it was in the actual situation and in the recorded video. The software can record the start and end time of each task using a straightforward technique. The programme enables the operator to see the video at various FPS (frames per second) and record the duration of each activity. In this investigation, the frame rate was fixed between 1.0 and 0.6 FPS for exact observation and time study data collection. With the use of software capabilities, the VA, NVA, and SVA timings were recorded (Fig. 15).

Table 1 (continued)

Task	Task description	Included resource					
		Main conv.	Robot	Right conv.	Left conv.	Right slider	Left slider
16	Left slider unloading the blue object to blue buffer						✔
17	Robot arm picking the yellow object from main conveyor	✔					
18	Left slider moves to its home position after unloading blue object		✔				
19	Robot arm placing yellow object to right conveyor		✔				
20	Right conveyor handling yellow object to right slider			✔			
21	Robot arm moves to its home position after placing yellow object		✔				

(continued)

The time study report for each task and section of the system was prepared independently for each resource/material. The data from the left, right, and front sections of the system were integrated to provide a thorough time study report of the system.

3.5.3 "Analyze" Phase of Six Sigma

During the "Measure" phase of Lean Six Sigma, a time study report was prepared that supplied the necessary information to identify problems, constraints, and associated resources. In the "Analyze" phase of Lean Six Sigma, this report on a time study was

Table 1 (continued)

Task	Task description	Included resource					
		Main conv.	Robot	Right conv.	Left conv.	Right slider	Left slider
22	Main conveyor handling green object to robot					✔	
23	Right slider transfers yellow object to yellow buffer					✔	
24	Right slider unloading the yellow object to yellow buffer					✔	
25	Robot arm picking the green object from main conveyor	✔					
26	Right slider moves to its home position after unloading yellow object		✔				
27	Robot arm placing green object to left conveyor		✔				
28	Left conveyor handling green object to left slider				✔		

(continued)

analyzed to identify cause factors. Obtaining the causal factors associated with each problem and limitation helps to identify the following:

- Exact moment for problems occurrence.
- The resource(s) and task(s) connected with the occurrence of issues and limitations.
- The resource/s and associated task/s in which the problems and limitations occurred.
- The reason behind the problems and limitation occurrence.

Table 1 (continued)

Task	Task description	Included resource					
		Main conv.	Robot	Right conv.	Left conv.	Right slider	Left slider
29	Robot arm moves to its home position after placing green object		✔				
30	Left slider transfers green object to green buffer						✔
31	Left slider unloading the green object to green buffer						✔
32	Left slider moves to its home position after unloading green object						✔

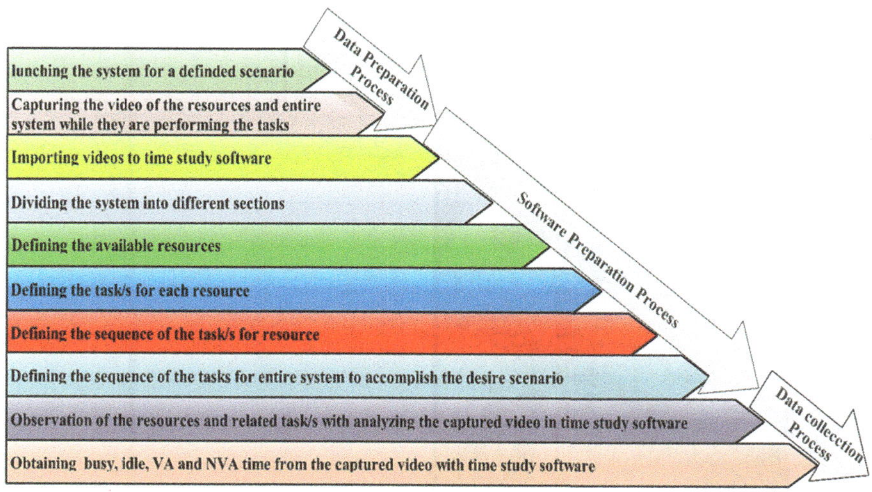

Fig. 13 Schematic of time study data collection procedure

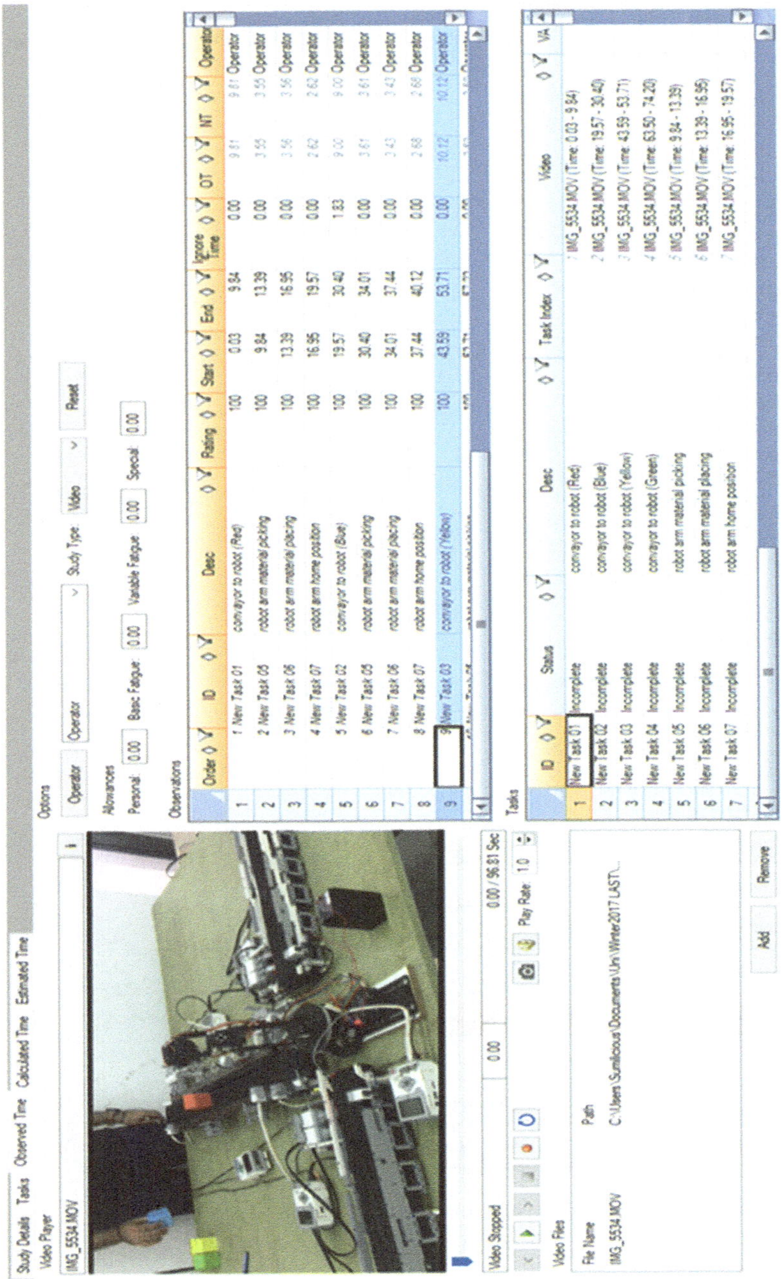

Fig. 14 Time study software GUI

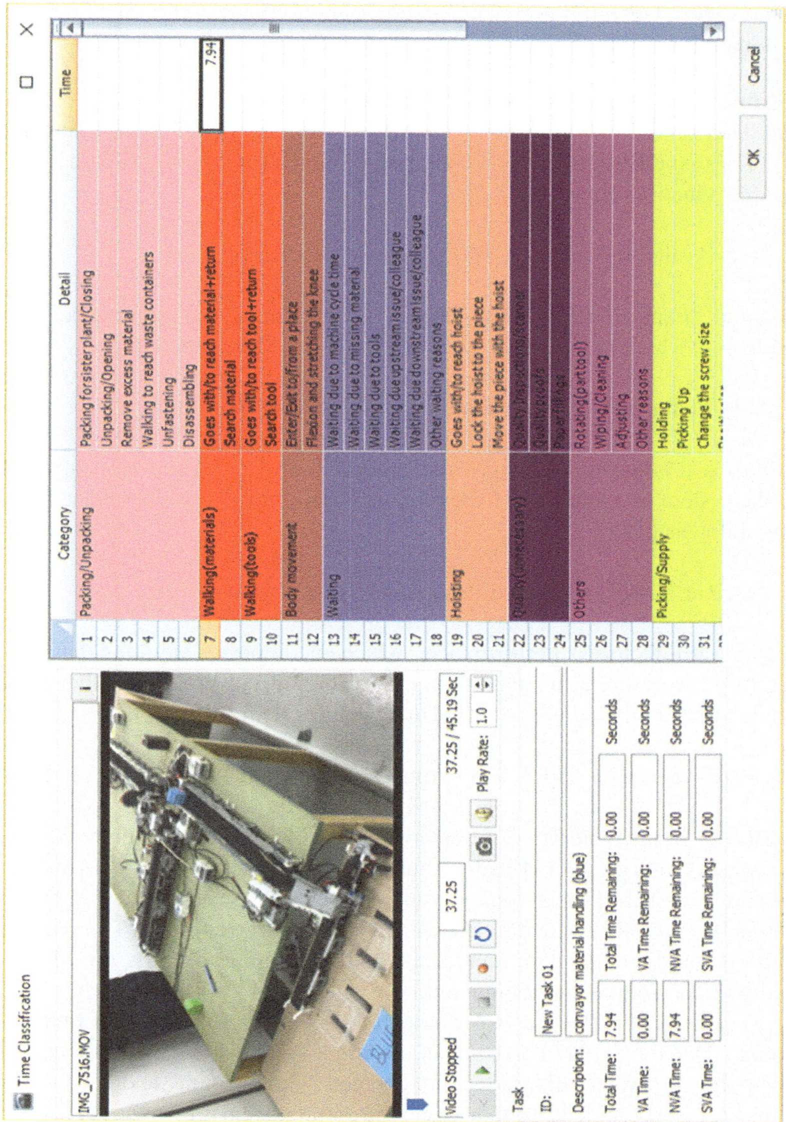

Fig. 15 Value added, non-value added and semi-value added times definite

These causal factors influence the system's performance through altering each resource and associated task/s. As a result, it was critical to use an accurate technique that included standards to evaluate the system's performance. This method must be capable of evaluating performance partially for each task, resource, and the entire system. Thus, Overall Equipment Effectiveness (OEE) was chosen as the most efficient method for evaluating the system's performance.

Overall Equipment Effectiveness and performance evaluation
In order to determine the OEE, each of the three OEE Factors had to be calculated separately. The first OEE factor was "Availability," which is the ratio of "Run Time" to "Planned Production Time" (5).

$$\text{Availability} = \text{Run Time}/\text{Planned Production Time} \tag{5}$$

To compute "Planned Production Time," "Cycle Time" was subtracted from scheduled losses, which in this study were referred to as "Idle time" (6).

$$\text{Planned Production Time} = \text{Cycle time} - \text{Idle time} \tag{6}$$

"Run time" is "Planned Production Time" minus "Stop Time" (Downtime), which is all time the system was scheduled to be running but was not due to unplanned Stops such as breakdowns (7).

$$\text{Run Time} = \text{Planned Production Time} - \text{Down Time} \tag{7}$$

The second OEE factor is "Performance," which encompasses anything that stops a resource from operating at its maximum pace. The ratio of product of the "Ideal Cycle Time" and "Total Count" to "Run Time" defines performance (8).

$$\text{Performance} = (\text{Ideal Cycle time} \times \text{Total Count})/\text{Run time} \tag{8}$$

In this study, "Ideal Cycle Time" refers to the theoretical minimum time required to complete task/s for a product in each resource. For example, the minimum amount of time a part must travel on the main conveyor before the robot arm can pick it up. This time has been theoretically determined based on the conveyor's length and speed. In addition, "Total Count" represents the total number of parts for which a task linked with a resource was completed.

Quality, the final OEE factor, was computed as the ratio of "Good Count" to "Total Count." Due to the characteristics of the target system and the accompanying production plan, all parts in this study were deemed "Good Count" and defect-free. However, as previously said, the primary function of the system is to identify products that have passed quality control (9).

$$\text{Quality} = \text{Good Count}/\text{Total Count} \tag{9}$$

3.5.4 "Improve" Phase of Lean Six Sigma

In this phase of Lean Six Sigma, all possible solutions to the problem, limitations, and associated causative factors established in the previous phase were evaluated. After implementing these changes, it was anticipated that system performance would increase. However, due to the nature of the target system, the implementation of these solutions with verifying their impact on system performance was crucial, as stated in Sect. 1. Therefore, validation of each solution to overcome the challenges and constraints and enhance the system's performance was necessary. To achieve this objective, discrete event simulation was used to validate the impact of the proposed solutions on the performance of each resource and the overall system.

The Simulation model to verify the solutions for improving the system performance

Important roles are played by Resources, procedures, Variables, and Attributes in the construction of the simulation model of the target system in Arena. After defining these characteristics, the algorithm for the simulation model was built using Arena's flowchart modelling technique. The Functional Specification of the system served as the model's road map to guarantee that it was developed correctly to handle the challenge. In addition, the simulation model requires data from the actual system that has to be acquired through time studies, work sampling, etc., such as the speed of the motions and distances between resources.

After creating a simulation model with the same characteristics as the actual system, it was necessary to collect simulation data to verify that the model's behaviours and functioning were identical to those of the actual system. In order to construct the simulation database, it was important to establish the type of data required and the source of this data.

The reliability of the simulation model is validated depending on the user's acceptance of data analysis. If the data is near enough to the real system data, it is reliable enough to be updated to improve system performance based on the found solutions. In this study, the model inputs are actual data from the real system, and simulation model data will be compared to the real system's time study results. In every instance, the existing system is modified to reflect the simulation model for ongoing analysis and operational decision-making.

3.5.5 "Control" Phase of Six Sigma

The primary objective of the "Control" phase of Lean Six Sigma was to preserve and sustain the benefits made during the "Improve" phase. In this phase, an exhaustive analysis was required to determine the solutions and associated elements that have the greatest impact on system performance. Categorizing the solutions and factors facilitates the prioritization of each solution and element. In this study, multiple options were outlined for each resource, and the three parameters with the greatest impact on system performance were determined to be the distance between resources, speed of motion for each resource, and system layout design. For each resource, a priority was assigned to these criteria based on their impact on the resource's performance and the overall system's performance.

4 Result and Discussion

4.1 Overview

The study's findings are presented and analyzed in this section in relation to the study's purpose and objectives. The primary objective was to establish the performance of the system following the introduction of an agent-based control architecture using Lean Six Sigma-DMAIC techniques, the time study method, and the Overall Equipment Effectiveness standard. Some difficulties and limits related with the target system were identified and some solutions were presented during the "analyze" phase of Lean Six Sigma. A simulation model of the system was developed in order to assess the effect of the proposed solution on system performance. The system performance was evaluated using simulation results and compared to the system performance prior to the deployment of the solutions. Among the offered solutions, those having the greatest impact on the system's performance were identified in order to build a priority list for each of them that would lead to the highest system performance.

4.2 Discrete Time Study Results for Each Phase of the System

To meet the requirements of the "Measure" phase of the Lean Six Sigma strategy, and based on the established phases and related resource/s for the target system, a separate time study was conducted for each step. Lean Six Sigma's "Analyze" step requires comprehensive time study data for the complete system, which was obtained by collecting the time study results from each phase. The discrete time study assisted in identifying the issues and constraints connected with each phase by addressing the corresponding resource(s) where the issue occurred.

4.2.1 Time Study Result of the Main Conveyor

The initial phase of the time study focused on the primary conveyor and the robot arm, which contain the specific activities listed in Table 2. The primary conveyor's Utilization Gantt chart is depicted in Fig. 16. The beginning of the time study was deemed to be the placement of the first object (red object) on the conveyor. When the motion sensor detected the first object, the conveyor began to move for 9.81 s, until the object reached the position of the colour sensor at the end of the main conveyor. The robot then picked up the red object once the colour sensor spotted it. The conveyor was inactive for 9.73 s as the robot arm picked up the object off the primary conveyor and placed it on the secondary conveyor. After the red object was placed on the right side-conveyor, the main conveyor was engaged and the blue object was moved for 9.04 s to a location where the robot arm could pick it up. The identical procedure was followed for the yellow and green objects. The main conveyor idle time before moving the yellow and green objects was 10.48 and 10.25 s respectively. In addition, the primary conveyor took 9.86 s to deliver the yellow object and 9.62 s to send the green object to the robot arm (Table 2).

Figure 16 and Table 2 depict the various busy and idle times associated with each object on the primary conveyor. Due to the use of the same mechanism (main conveyor) and handling technique, it was anticipated that identical data would be seen. The lowest recorded time was associated with the blue object, indicating that the main conveyor delivered this object to the robot faster than the order objects.

Table 2 Main conveyor time study data

Task description	Start (s)	End (s)	Busy time (s)	Idle time (s)
Main conveyor handling red object to robot	0	9.81	9.81	9.73
Main conveyor handling blue object to robot	19.54	28.58	9.04	10.48
Main conveyor handling yellow object to robot	39.06	48.92	9.86	10.25
Main conveyor handling green object to robot	59.17	68.79	9.62	–

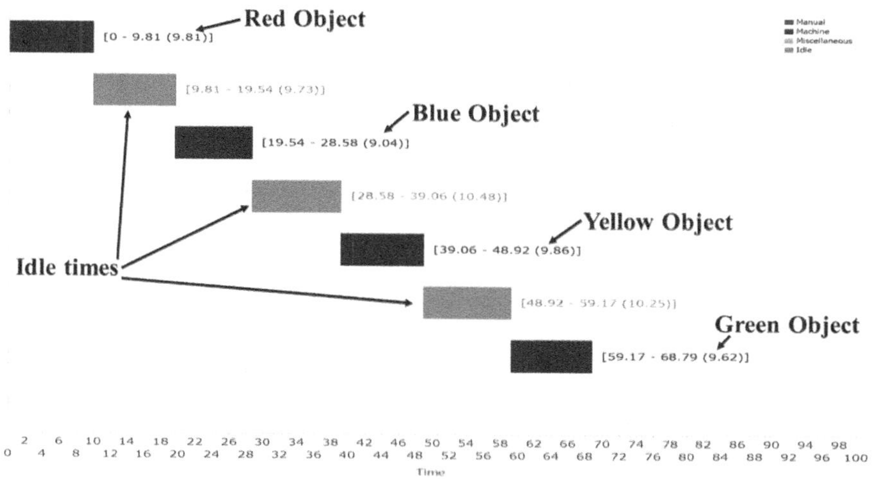

Fig. 16 Main conveyor utilization report

Identified Problem and Limitation for the Main Conveyor by Time Study and Possible Solutions

Following the "Analyze" phase of Lean Six Sigma techniques and applying the time study report related with the primary conveyor, all problems, limitations, and potential solutions for this resource have been discovered. All discovered issues with the primary conveyor were a result of the interaction between the control software in slave agent 1 and its interaction with slave agent 6.

The busy time difference induced by the predefined detection range of sensors utilized to activate the conveyor in slave agent 1. By modifying the detection range, sensors may detect objects and send the signal to the slave agent 1 faster, regardless of how the components are arranged on the conveyor.

The variance in idle times was primarily determined by the performance of the colour sensor and robot, respectively. By modifying the light intensity detection range for the colour sensor on slave agent 6, the colour of a part can be detected more quickly and the slave agent can activate the robot more quickly, regardless of where the part is detected within the colour sensor's detection region.

Slave agent 1 could not be engaged prior to receiving the signal from slave agent 6 regarding the completion of the robot arm task. In order to reduce idle time, it was necessary to adjust the speed of the robot's movements, as it affected the idle time of the conveyor.

In general, increasing the speed of motion and decreasing the distance between the resources in an effort to reduce the time is possible by expanding the power range of the available motors associated with the resource, which could be accomplished by modifying the agent software on slave agents 1 and 6. On the other hand, altering the layout design of the primary conveyor and the robot arm will reduce the duration of material handling between these two resources. For instance, by positioning the colour sensor (sensor 4) closer to sensor 2, the slave agent's ability to assess the signal and decide whether to move the part to the left or right side-conveyor based on the colour is enhanced.

4.2.2 Time Study Result of the Robot Arm

The operation of the robot arm was consistent, and the values for picking and positioning each object were practically the same. The start, end, and total duration of the robot arm tasks are displayed in Table 3. Except for the blue object, Table 3 observed that the procedure was constant. The table also indicates that the idle period for robots corresponds to the busy time for the primary conveyor. It indicates that the robot was in a mode of inactivity while the main conveyor moved the parts.

Table 3 Robot arm time study report

Task description	Start (s)	End (s)	Busy time (s)	Idle Time (s)
Red object picking	9.843	13.36	3.55	
Red object placing	13.36	16.92	3.56	
Robot arm home position	16.92	19.54	2.62	
Busy and Idle time for red object			**9.73**	**9.81**
Blue object picking	28.58	32.46	3.88	
Blue object placing	32.46	36.28	3.82	
Robot arm home position	36.28	39.06	2.78	
Busy and Idle time for blue object			**10.48**	**9.04**
Yellow object picking	48.92	52.97	4.05	
Yellow object placing	52.94	56.60	3.36	
Robot arm home position	56.60	59.17	2.57	
Busy and Idle time for yellow object			**9.98**	**9.84**
Green object picking	68.79	72.63	3.84	
Green object placing	72.63	76.23	3.60	
Robot arm home position	76.23	78.73	2.50	
Busy and Idle time for green object			9.94	**9.62**

Identified Problems and Limitations for Robot Arm by Time Study and Possible Solutions

The varying amount of time required for the robot to move the blue part revealed that the colour sensor had trouble distinguishing the object and the colour blue in a timely manner. Therefore, the agent software, including the definition of the light intensity detection range for this sensor in slave agent 6, must be adjusted in order to accurately recognize objects of various hues. Figure 17 demonstrates that the robot arm's idle times rely on the primary conveyor process. As previously stated, minimizing the idle time of the robot arm can be accomplished by boosting the process efficiency of the primary conveyor.

4.2.3 Time Study Result of the Right Side-Conveyor

Table 4 depicts the entire idle time of 50.29 s required to transport the red and yellow objects from the right side-conveyor to the right slider unit. The right side-conveyor was idle for 16.92 s to accept the red object, but 33.37 s to receive the yellow one. The right side-conveyor was in idle mode for 22.15 s while waiting for the completion of the entire process. This conveyor's idle time is a combination of the time it should be in idle mode and the time it must wait for the main conveyor and robot to complete their respective tasks. For instance, the idle time prior to moving the red object is 16.92 s, which is the sum of 9.81 s as a busy time for the main conveyor to move

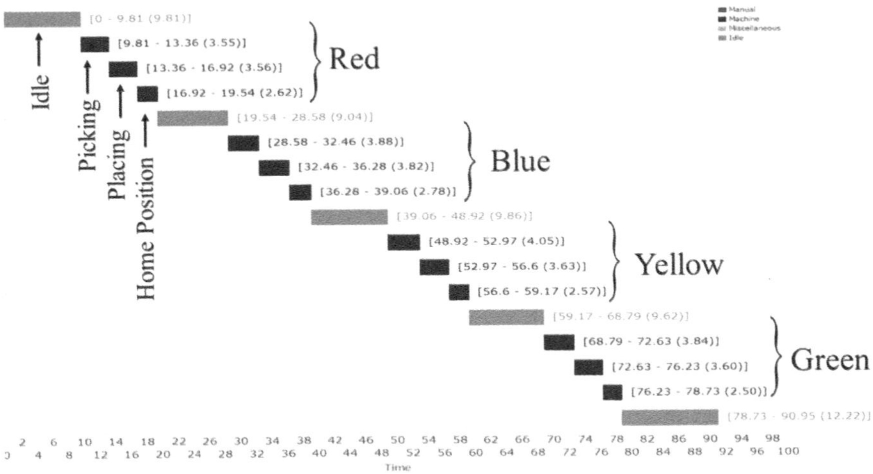

Fig. 17 Robot arm utilization report

this object and 3.55 and 3.56 s for the robot arm to remove the part from the main conveyor and set it on the right side-conveyor (Fig. 18).

Identified Problem and Limitation for Right Side-Conveyor by Time Study and Possible Solutions

Table 4 demonstrates that the busy time for the side-conveyor is smaller than the time spent as idle or non-value added time. The conveyor's performance is enhanced as a result of a reduction in idle time. Notably, this resource served as a link between the main conveyor and the right slider unit, with the performance of the resources in between affecting the side conveyor's performance. Therefore, enhancing the

Table 4 Right side conveyors time study data

Task description	Start (S)	End (s)	Busy time (s)	Idle time (s)
Right side-conveyor idle time before moving red object				16.92
Right side-conveyor for the red object	16.92	26.14	9.22	
Right side-conveyor idle time before moving yellow object				33.37
Right side-conveyor for yellow object	59.51	68.81	9.30	
Overall busy and idle time for right side-conveyor			**18.52**	**50.29**
Right side-conveyor idle time after moving yellow object				22.15

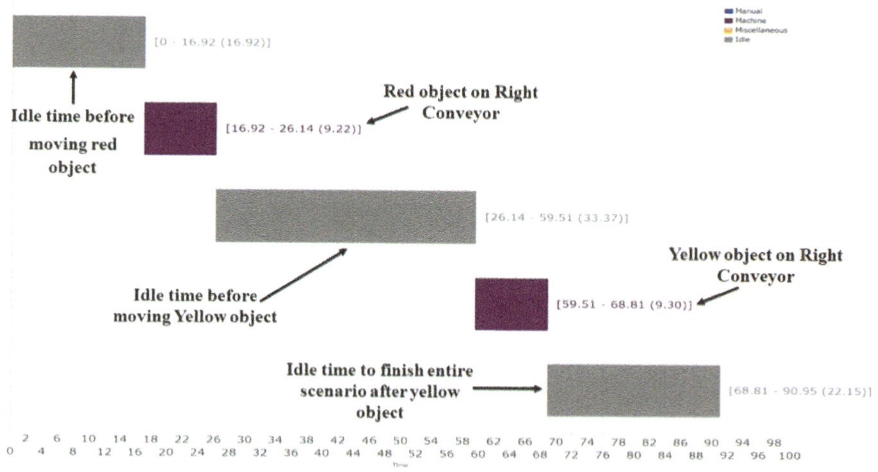

Fig. 18 Right side-conveyor utilization report

performance of the primary conveyor, the robot, and the right slider unit will reduce the right side-conveyor's downtime. However, the software agents in slave agents 1, 2, 4, and 6 must be updated in terms of their control logic and communication with master agents. In addition, modifying the layout design to enhance the entrance and exit point (the point where the right side-conveyor connects to the slider) may be a viable solution.

4.2.4 Time Study Result of the Right Slider Unit

Figure 19 demonstrates that the right slider was in idle mode throughout the experiment, and according to the findings of the time study, idle time is greater than busy time for this resource. This resource has three primary responsibilities: transferring the object to the destination buffer (Buffer for red and objects), unloading the object from the buffers, and returning the object to its starting place (home). The total amount of time spent on these activities has been categorized as busy time. Due to the differing system behaviour for moving the red and yellow objects, separate time studies were conducted for each step. This slider was in the idle state for 26.14 s before moving the red item, and it took 12.49 s to move the object to the buffer and return to the home position (Table 5).

Since the distance between the buffers associated with red and yellow items differed, the amount of time required to move these objects also differed. The unloading task of sliders is the task associated with a mechanism within the slider that unloads objects to the buffer. Red and yellow objects have respective busy times of 4.50 and 4.32 s for unloading, which is practically the same. In addition, the red object takes 3.97 s to move to its associated buffer, while the yellow object takes

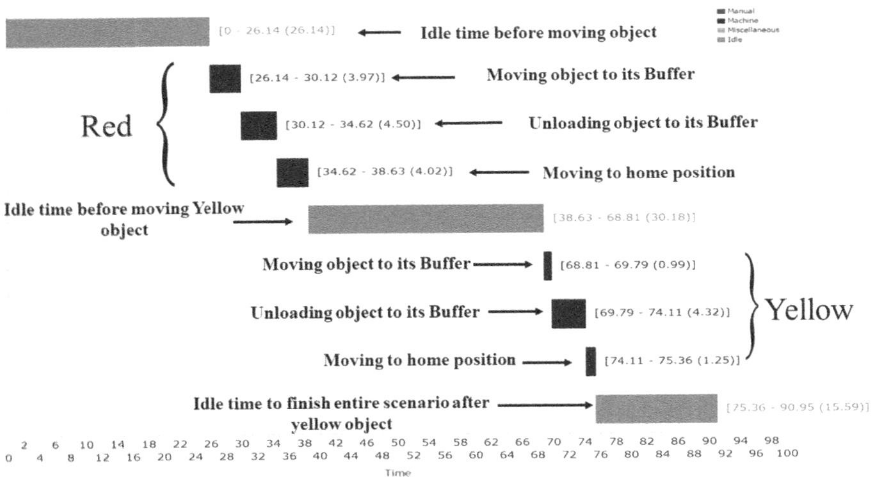

Fig. 19 Right slider utilization report

Table 5 Right sliders time study data

Task description	Start (S)	End (s)	Busy time (s)	Idle time (s)
Right slider idle time before moving red object				26.14
Right slider to red buffer	26.14	30.12	3.97	
Right slider unloading red	30.12	34.62	4.50	
Slider home position after unloading red	34.62	38.63	4.02	
Overall busy and idle time for right side-conveyor to move red object			**12.49**	**26.14**
Right slider Idle time before moving yellow object				30.18
Right slider to yellow buffer	68.81	69.79	0.99	
Right slider unloading yellow	69.79	74.11	4.32	
Slider home position after unloading yellow	74.11	75.36	1.25	
Overall busy and idle time for right side-conveyor to move yellow object			**6.56**	**30.18**
Right slider Idle time after moving red object				15.59

0.99 s. Due to the distance between buffers, the time required for the slider to return to its initial position varies substantially (Table 5).

Identified Problem and Limitation for Right Slider by Time Study and Possible Solutions

The resource's idle time is dependent on the performance of the preceding resources (main conveyor, robot arm, and right side-conveyor) that delivered the object to this slider. To reduce idle time for the slider unit, the performance of the aforementioned resources must be enhanced. To achieve this objective, the agent software in Slave agents 1, 2, and 6 must be modified and the solution proposed in the preceding section must be implemented on these resources.

4.2.5 Time Study Result of the Left Side-Conveyor and Left Slider Unit

Since the functioning and hardware design of the left side-conveyor and left slider unit are identical to those of the right side-conveyor and right slider, the identical analysis applies to their time study results (Figs. 20 and 21). The total idle time for the left side-conveyor is 68.29 s, while the total idle time for the left slider unit is 46.57 s for blue objects and 24.60 s for green objects. (Tables 6 and 7) The busy time associated with the left side-conveyor for transporting blue and green objects is practically identical: 7.94 and 7.88 s, respectively. The result of the time study for the left slider follows the same trend as the right slider. It indicates that the left slider for moving the blue object has almost the same value as the right slider for moving the red object and the yellow slider for moving the green object. Also, travelling back to the home position after unloading the objects took almost the same amount of time.

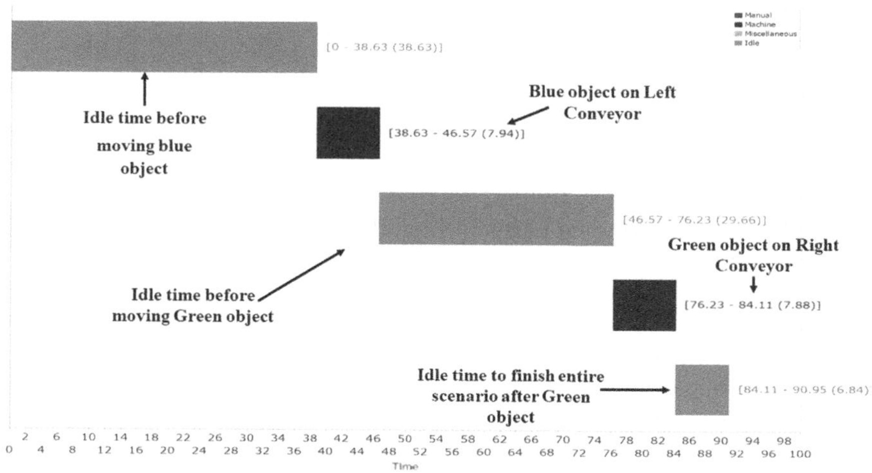

Fig. 20 Left side-conveyor utilization report

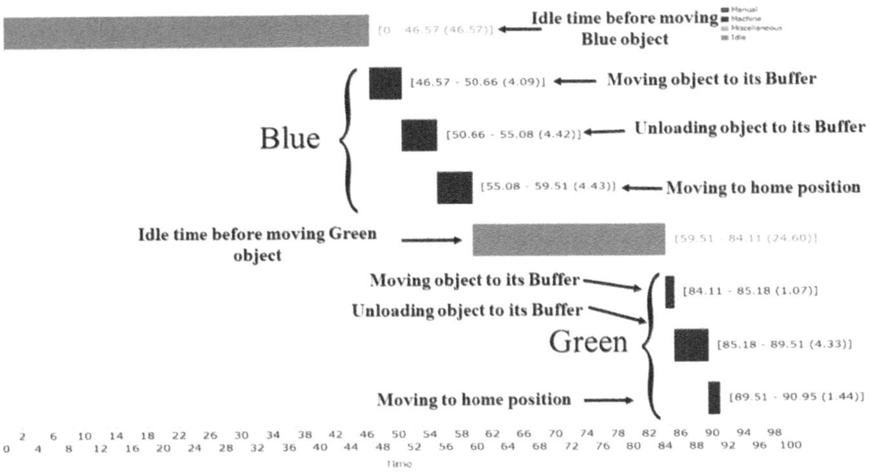

Fig. 21 Left slider utilization report

Table 6 Left side-conveyor time study data

Task description	Start (S)	End (s)	Busy time (s)	Idle time (s)
Left side-conveyor Idle time before moving blue object				38.63
Left side-conveyor for blue object	38.63	46.57	7.94	
Left side-conveyor Idle time before moving Green object				29.66
Left side-conveyor for Green Object	76.23	84.11	7.88	
Overall busy and idle time for left side-conveyor			**15.82**	**68.29**
Left side-conveyor Idle time after moving Green object				6.84

4.2.6 Comparison Between Left and Right Side-Conveyor Time Study Results

The right side-conveyor (assigned for Red and Yellow objects) takes 2.7 s longer to move the objects to the slider than the left side-conveyor (given for Green and Blue products). After conducting a precise time study specifically for this segment and analyzing the data, it was determined that the conveyors were created with differing lengths (Right Conveyor 10 cm Longer than Left) but the times recorded are accurate. As a result of the greater length of the right conveyor, this substantial difference is justified.

Table 7 Left slider time study data

Task description	Start (S)	End (s)	Busy time (s)	Idle time (s)
Left slider Idle time before moving blue object				46.57
Left slider to blue buffer	46.57	50.66	4.09	
Left slider unloading blue	50.66	55.08	4.42	
Slider home position after unloading blue	55.08	59.51	4.43	
Overall busy and idle time for left side-conveyor to move the blue object			**12.94**	**46.57**
Left slider Idle time before moving green object				24.60
Left slider to green buffer	84.11	85.18	1.07	
Left slider unloading green	85.18	89.51	4.33	
Slider home position after unloading green	86.51	90.95	1.44	
Overall busy and idle time for left side-conveyor to move the green object			**6.84**	**24.60**

4.2.7 Comparison Between Left and Right Slider Unit Time Study Results

The results of the time study indicate that the slider behaves similarly while moving red and blue objects as well as green and yellow ones. Since the distances to buffers for sliders used to move parts and return to the home position are identical, the busy time for blue and red objects, as well as yellow and green objects, are identical. Therefore, the little variation in results was due to the light intensity range of colour sensor detection and its time delay, which were defined in slave agents 4 and 5.

4.2.8 Overall Time Study Result of the Entire System

The following Gantt chart (Fig. 22) and Table 5 depict the time study report for the entire system, illustrating the busy and idle time of each resource as well as the behaviour of two or more resources working concurrently. The Gantt chart was built using the table of task dependencies (Tables 6, 7, 8, 9 and 10).

4.3 Utilization Rate Report to Identify the Resources with More Idle Time

Since acquiring techniques to optimize and improve the system was part of the Lean Six Sigma objective, the difficulties and limitations connected with each resource in the target system were acquired and reported in Table 11. Utilization rate was chosen

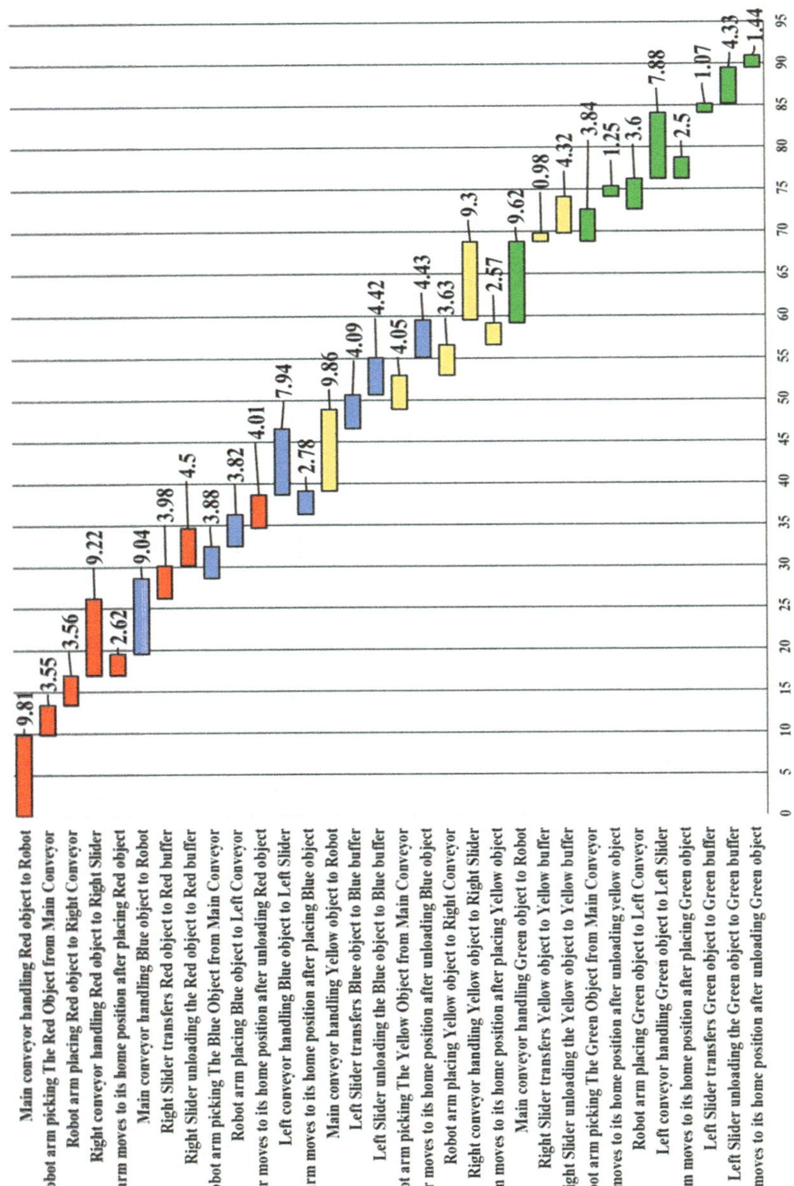

Fig. 22 Overall time study report

Table 8 Detailed time result of the tasks

Task no	Task description	Start time	End time	Busy time
1	Main conveyor handling red object to robot	0	9.81	9.81
2	Robot arm picking the red object from main conveyor	9.81	13.36	3.55
3	Robot arm placing red object to right conveyor	13.36	16.92	3.56
4	Right conveyor handling red object to right slider	16.92	26.14	9.22
5	Robot arm moves to its home position after placing red object	16.92	19.54	2.62
6	Main conveyor handling blue object to robot	19.54	28.58	9.04
7	Right slider transfers red object to red buffer	26.14	30.12	3.98
8	Right slider unloading the red object to red buffer	30.12	34.62	4.5
9	Robot arm picking the blue object from main conveyor	28.58	32.46	3.88
10	Robot arm placing blue object to left conveyor	32.46	36.28	3.82
11	Right slider moves to its home position after unloading red object	34.62	38.63	4.01
12	Left conveyor handling blue object to left slider	38.63	46.57	7.94
13	Robot arm moves to its home position after placing blue object	36.28	39.06	2.78
14	Main conveyor handling yellow object to robot	39.06	48.92	9.86
15	Left slider transfers blue object to blue buffer	46.57	50.66	4.09
16	Left slider unloading the blue object to blue buffer	50.66	55.08	4.42
17	Robot arm picking the yellow object from main conveyor	48.92	52.97	4.05
18	Left slider moves to its home position after unloading blue object	55.08	59.51	4.43
19	Robot arm placing yellow object to right conveyor	52.97	56.6	3.63
20	Right conveyor handling yellow object to right slider	59.51	68.81	9.3
21	Robot arm moves to its home position after placing yellow object	56.6	59.17	2.57
22	Main conveyor handling green object to robot	59.17	68.79	9.62
23	Right slider transfers yellow object to yellow buffer	68.81	69.79	0.98

(continued)

Table 8 (continued)

Task no	Task description	Start time	End time	Busy time
24	Right slider unloading the yellow object to yellow buffer	69.79	74.11	4.32
25	Robot arm picking the green object from main conveyor	68.79	72.63	3.84
26	Right slider moves to its home position after unloading yellow object	74.11	75.36	1.25
27	Robot arm placing green object to left conveyor	72.63	76.23	3.6
28	Left conveyor handling green object to left slider	76.23	84.11	7.88
29	Robot arm moves to its home position after placing green object	76.23	78.73	2.5
30	Left slider transfers green object to green buffer	84.11	85.18	1.07
31	Left slider unloading the green object to green buffer	85.18	89.51	4.33
32	Left slider moves to its home position after unloading green object	89.51	90.95	1.44

to determine which resources had the most Idle time. Table 10 displays the utilization rate for each particular resource. The resources with the lowest utilization rate were highlighted, and the issues that led to increased idle time for those resources were identified. According to OEE, to determine the utilization rate, Busy and Idle time must be combined together to obtain production time, and production time divided by Busy time yields the respective utilization of each resource.

The area under the Gantt chart in Fig. 22 reveals that the complete scenario requires 90.95 s. The sum of the Busy times of the tasks in Table 8 is 151.89 s, which is longer than the total duration of the process. Tables 8 and 9 illustrate that the majority of tasks are performed concurrently or simultaneously. It may be stated that when Idle time decreases, the ability to run tasks concurrently grows. It is crucial to recognize that reducing Idle time without reducing Busy time is impossible. Consequently, the control architecture of slave agents must be adjusted so as to maximize idle times in consideration of busy times. For instance, if the robot's busy time lowers, the conveyor's idle time to load the next object will also decrease. The overall Idle time might be optimized by modifying the system's hardware, software (control software in agents), and layout design; as a result, the system's performance would be enhanced. Table 11 provides a summary of the observed difficulties and potential solutions for improving timing using the time study technique.

Table 9 Task dependencies of the system

Task no	Task description	Predecessor task	Related resources
1	Main conveyor handling red object to robot	1	Main conv
2	Robot arm picking the red object from main conveyor	2	Robot arm
3	Robot arm placing red object to right conveyor	3	Robot arm
4	Right conveyor handling red object to right slider	4	Right conv
5	Robot arm moves to its home position after placing red object	4	Robot arm
6	Main conveyor handling blue object to robot	5	Right slider
7	Right slider transfers red object to red buffer	6	Right slider
8	Right slider unloading the red object to red buffer	8	Right slider
9	Robot arm picking the blue object from main conveyor	7	Main conv
10	Robot arm placing blue object to left conveyor	9	Robot arm
11	Right slider moves to its home position after unloading red object	10	Robot arm
12	Left conveyor handling blue object to left slider	12	Left conv
13	Robot arm moves to its home position after placing blue object	11	Robot arm
14	Main conveyor handling yellow object to robot	13	Left slider
15	Left slider transfers blue object to blue buffer	14	Left slider
16	Left slider unloading the blue object to blue buffer	16	Left slider
17	Robot arm picking the yellow object from main conveyor	15	Main conv
18	Left slider moves to its home position after unloading blue object	18	Robot arm
19	Robot arm placing yellow object to right conveyor	17	Robot arm
20	Right conveyor handling yellow object to right slider	21	Right conv
21	Robot arm moves to its home position after placing yellow object	19	Robot arm
22	Main conveyor handling green object to robot	20	Right slider
23	Right slider transfers yellow object to yellow buffer	23	Right slider

(continued)

Table 9 (continued)

Task no	Task description	Predecessor task	Related resources
24	Right slider unloading the yellow object to yellow buffer	24	Right slider
25	Robot arm picking the green object from main conveyor	22	Main conv
26	Right slider moves to its home position after unloading yellow object	26	Robot arm
27	Robot arm placing green object to left conveyor	25	Robot arm
28	Left conveyor handling green object to left slider	27	Left conv
29	Robot arm moves to its home position after placing green object	27	Robot arm
30	Left slider transfers green object to green buffer	28	Left slider
31	Left slider unloading the green object to green buffer	29	Left slider
32	Left slider moves to its home position after unloading green object	30	Left slider

Table 10 System utilization report

Resource	Busy time (s)	Idle time (s)	Cycle time (s)	Utilization (%)
Main conveyor	40.40	30.46	70.86	57.01
Left conveyor	15.82	75.13	90.95	21.05
Right conveyor	18.52	72.44	90.96	20.36
Robot arm	34.34	50.55	80.89	42.45
Left slider	19.04	71.17	90.21	21.10
Right slider	19.78	71.91	91.69	21.57

4.4 Simulation Result of the System by Arena

For Arena to build a realistic model and achieve behaviour similar to that of the actual system, busy time was chosen as the aim. Obtaining the busy time for each resource and its associated task(s), the required velocity of the resources, and the distance between each section of the actual system. For this reason, five participants with a thorough understanding of the system and its operations were enlisted to measure and collect the necessary data using instruments with the highest accuracy and precision.

Using a motion sensor, the velocity of objects and resources has been detected. In this technology, an electrostatic transducer on the Motion Sensor emits ultrasonic pulses with a frequency of approximately 49 kHz. To measure velocity, the sensor detects the time between the trigger rising edge and the echo rising edge. For the

Table 11 Detected problems and limitations and proposed solutions

Resource	Detected problems and limitation	Reason	Solution
Main conveyor	Different busy time for moving the objects	Detection range of the motion sensors	Modification of the detection range (range for sensor 1, 2 and 3 on slave agent 1)
	Different idle time before moving the objects	Detection range of the colour sensor	Modification of light intensity detection range (range for sensor 4 on slave agent 6)
		Robot performance dependency	Optimizing the speed of the robot motion by modifying the agent software on slave agent 6
			Modifying the layout design to decrease the distance between the sensors on main conveyor
Robot arm	The different busy time for robot to move the blue part	Detection range of the colour sensor	Modification of light intensity detection range (range for sensor 4 on slave agent 6)
	Long idle time before moving the objects	Main conveyor performance dependency	Optimizing the speed of the conveyor motion by modifying the agent software on slave agent 1
Right side-conveyor	Low utilization rate (long idle time low busy time)	Main conveyor and robot arm and right slider performance dependency	Improving the performance of main conveyor, robot and right slider by modifying the agent software on slave agent 1, 2 and 4
			Improving the communication between main conveyor, robot and right slider by modifying the interaction protocols in agent software on slave agent 1, 2 and 4 and master agent

(continued)

Table 11 (continued)

Resource	Detected problems and limitation	Reason	Solution
			Modifying the layout design to optimize the place of exit point on right side-conveyor
	Different busy time in comparison with left side-conveyor	Different length of the conveyor	Modifying the length of the conveyor or increasing the speed of conveyor motion by modifying the slave agent 2
Right slider	Low utilization rate (long idle time low busy time)	Main conveyor and robot arm and right side-conveyor performance dependency	Improving the performance of main conveyor, robot and right slider by modifying the agent software on slave agent 1, 2 and 4
			Improving the communication between main conveyor, robot and right slider by modifying the interaction protocols in agent software on slave agent 1, 2 and 4 and master agent
			Modifying the layout design to optimize the place of exit point on right side-conveyor
	Different idle times	Different distance between home position and buffers	Modifying the layout design to change the buffers place to be near to each other

most precise findings, the distance between the resources and the displacement of the objects was measured with a laser distance metre. Each participant measured independently and was instructed to calculate the time based on the measured velocity and distance (Table 12).

The next step in developing an accurate simulation model is to model the agent-based control architecture with master–slave mechanism. Figure 13 illustrates the entire simulation model control architecture described in Sect. 3. In order for a simulation model to replicate the real system's behaviour as closely as possible, it was necessary to consider every hardware, software, and layout design specification present in the real system. The model was therefore divided into the same districts as

Table 12 Measured distance and speed by participants

From	To	Among	Resource	Distance (free Unit)	Speed (distance unit per Second)
Entrance/Sensor 1	Sensor 2	Main conv.	Main conv.	53	9.3
Sensor 2	Sensor 3 and 4	Main conv.	Main conv.	12	3.35
Sensor 3 and 4	Sensor 5	Main conv. and right conv	Robot arm	34	12.9
Sensor 3 and 4	Sensor 6	Main conv. and left conv	Robot arm	29	10.4
Sensor 5	Sensor 7/Right slider entrance	Right conv. and right slider	Right conv.	85	9
Sensor 7/Right slider entrance	Red buffer	Right slider	Right slider	8	8
Sensor 7/Right slider entrance	Yellow buffer	Right slider	Right slider	39	10
Sensor 8/Left slider entrance	Green buffer	Left slider	Left slider	7	8
Sensor 8/Left slider entrance	Blue buffer	Left slider	Left slider	39	10
Sensor 6	Sensor 8/Left slider entrance	Left conv. and left slider	Left conv.	75	9
Right slider	Red buffer	Right slider and red buffer	Right slider	1	2
Right slider	Yellow buffer	Right slider and yellow buffer	Right slider	1	3.13
Left slider	Blue buffer	Left slider and blue buffer	Left slider	1	3.13
Left slider	Green buffer	Left slider and green buffer	Left slider	1	2.39

the actual system, including the system's entrance and the point at which parts were loaded into the system, the main conveyor, Right and Left Conveyor, Right and Left Slider units, and the robotic arm (Fig. 23).

To develop the simulation model in the Arena, the genuine system's exact attributes were required. By measuring the distances between the resources and velocities of the motions in the same manner as the real system and incorporating

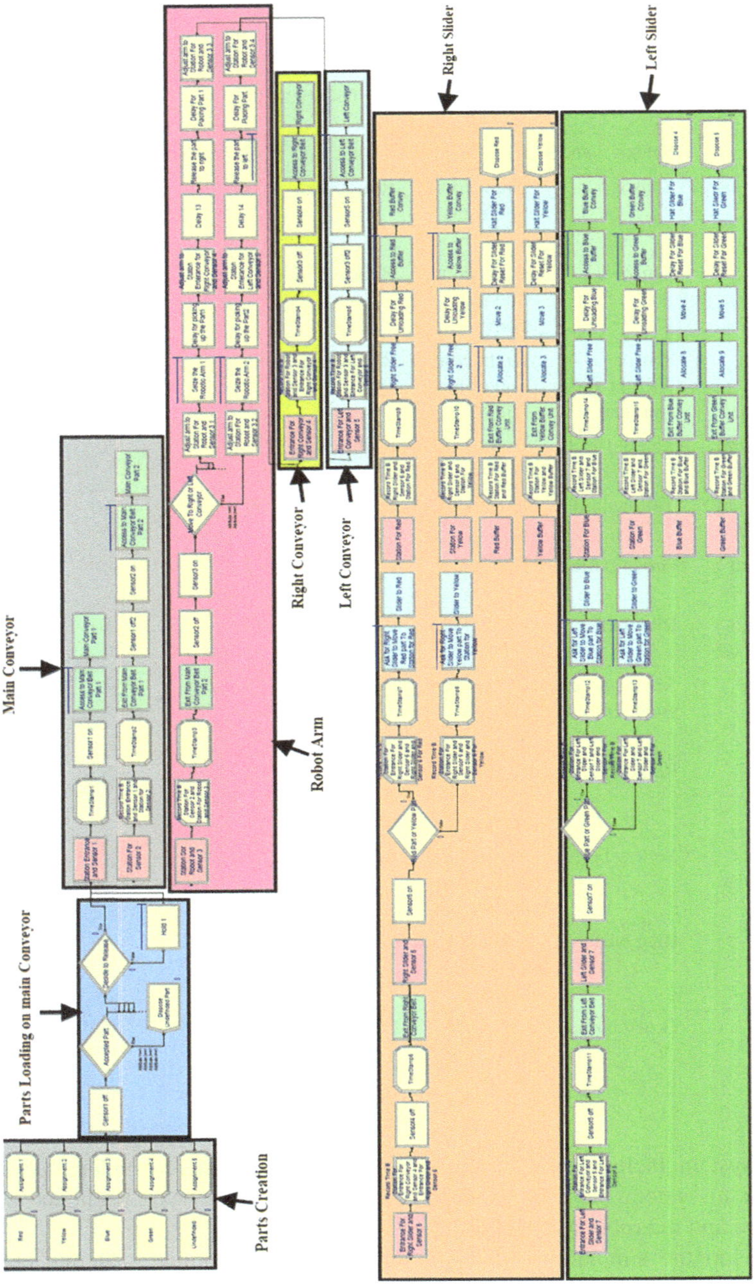

Fig. 23 Arena simulation model

them into the model, the simulation resulted in a time that was identical to the real system's time study result. This simulation model was accurate enough to evaluate the effects of modifying the proposed solution in Table 11 and evaluating the amended model's effects on the system. In the simulation model, an accurate average of the participant's measurements was applied, and the results were substantially identical to those of the time study of the actual system. This result is shown in Table 13. In addition, visualization in the Arena simulation provided a graphical depiction and animation that improved comprehension of the model's behaviour. This visualization assisted in gaining an understanding of the implications of suggested system modifications (Fig. 24).

Table 13 displays the absolute time difference of 0.5 s between simulation model and actual system busy time (Fig. 25). So that the simulation model verifies and validates the functional specifications and any modifications or adjustments can be applied with confidence.

4.5 OEE Analysis Before Optimization

Following the design and implementation of the suggested agent-based control architecture on the target material handling system and the completion of a comprehensive time study, the overall system performance was determined by analyzing the outcome. To achieve this objective, the OEE standard was studied and implemented. Tables 2, 3, 4, 5, 6 and 7 were applied to determine the partial and whole system performance, and Table 14 was created accordingly.

OEE percentages are determined separately for each system component (each resource). This percentage provides an accurate representation of the effectiveness of the resource. This percentage facilitates the identification and tracking of low-performing resources and the evaluation of their performance following any modifications (solutions) (Fig. 26).

First component entirely dependent on resource downtime, idle time, and scheduled production time is resource availability. Figure 27 demonstrates that the availability of all resources exceeds 90%, with the exception of the left Slider for the Green item. It indicates that this resource's downtime and breakdowns were not as efficient as those of other resources. The primary conveyor has a downtime of 2.3 s, which is a result of the primary conveyor's incapacity to accommodate the movement of different-shaped products. In addition, the primary conveyor has an idle period of 30.46 s due to the performance of the robot. The robot has 99% availability due to its total Idle time and downtime of 50.49 s out of 90.95 s, which is the lowest value among comparable resources. In availability percentage analysis, the slider's functionality is separated into several tasks. The rationale for this distinction is that sliders transfer objects to buffers at varying distances based on their behaviour. The reason for the lower availability % of sliders compared to other resources is the instability of the tray in slider units when they are performing and their speed of motion.

Table 13 Time difference between the real system and simulation model busy times

		Busy time (s)		
Task no	Task description	Real system	Average of simulation result using participants observation	Absolut deference
1	Main conveyor handling red object to robot	9.81	9.50	0.31
2	Robot arm picking the red object from main conveyor	3.55	3.84	0.29
3	Robot arm placing red object to right conveyor	3.56	3.66	0.10
4	Right conveyor handling red object to right slider	9.22	9.40	0.18
5	Robot arm moves to its home position after placing red object	2.62	2.50	0.12
6	Main conveyor handling blue object to robot	9.04	9.50	0.46
7	Right slider transfers red object to red buffer	3.98	3.90	0.08
8	Right slider unloading the red object to red buffer	4.5	4.50	0.00
9	Robot arm picking the blue object from main conveyor	3.88	3.55	0.33
10	Robot arm placing blue object to left conveyor	3.82	3.66	0.16
11	Right slider moves to its home position after unloading red object	4.01	4.08	0.07
12	Left conveyor handling blue object to left slider	7.94	8.33	0.39
13	Robot arm moves to its home position after placing blue object	2.78	2.50	0.28
14	Main conveyor handling yellow object to robot	9.86	9.50	0.36
15	Left slider transfers blue object to blue buffer	4.09	4.09	0.00
16	Left slider unloading the blue object to blue buffer	4.42	4.40	0.02

(continued)

Table 13 (continued)

		Busy time (s)		
Task no	Task description	Real system	Average of simulation result using participants observation	Absolut deference
17	Robot arm picking the yellow object from main conveyor	4.05	3.84	0.21
18	Left slider moves to its home position after unloading blue object	4.43	4.43	0.00
19	Robot arm placing yellow object to right conveyor	3.63	3.66	0.03
20	Right conveyor handling yellow object to right slider	9.3	9.40	0.10
21	Robot arm moves to its home position after placing yellow object	2.57	2.50	0.07
22	Main conveyor handling green object to robot	9.62	9.50	0.12
23	Right slider transfers yellow object to yellow buffer	0.98	0.99	0.01
24	Right slider unloading the yellow object to yellow buffer	4.32	4.32	0.00
25	Robot arm picking the green object from main conveyor	3.84	3.55	0.29
26	Right slider moves to its home position after unloading yellow object	1.25	1.25	0.00
27	Robot arm placing green object to left conveyor	3.6	3.66	0.06
28	Left conveyor handling green object to left slider	7.88	8.33	0.45
29	Robot arm moves to its home position after placing green object	2.5	2.50	0.00

(continued)

Table 13 (continued)

		Busy time (s)		
Task no	Task description	Real system	Average of simulation result using participants observation	Absolut deference
30	Left slider transfers green object to green buffer	1.07	0.71	0.36
31	Left slider unloading the green object to green buffer	4.33	4.32	0.01
32	Left slider moves to its home position after unloading green object	1.44	1.44	0.00

In addition, sliders are the final system resources and have a significant amount of observed idle time.

Figure 28 depicts the performance percentage of the system's available resources. As stated in Sect. 1, the performance of a resource is a ratio between the ideal cycle time and the actual run time for a certain number of products. If the percentage of performance is approaching 100, it indicates that the resource has less run time and more Ideal cycle time and is operating as quickly as feasible.

The performance of sliders for moving yellow and green objects (moving the objects to a buffer close to the slider's home position) is almost identical at 83.19 and 83.33%, respectively. The performance of the Red and Blue object-moving sliders is 91.67%. Due to the unanticipated length disparity, the percentage of left and right side conveyors' performance varies (right conveyor 10 cm longer than the left one). The performance of the right side-conveyor is 94.44%, while the performance of the left side-conveyor is 90.32%. The result demonstrates that although the right conveyor is longer, its performance is superior. This contrast demonstrates that the performance of the right conveyor is superior due to a faster object-moving speed and less downtime with a greater length. In addition, the performance of the side conveyors is dependent on the Ideal Cycle Time, the Run Time, and the total number of counted products. Total products counted and Ideal cycle time are deemed identical for both conveyors. Therefore, Run Time should be the sole cause of the cited problem. Run time is heavily impacted by Idle time and downtime, both of which have less value for the Right conveyor. It is important to realize that idle and downtimes influence other resources in addition to the conveyors themselves. It indicates that, despite the fact that the right conveyor is 10 cm longer than the left one, the correct values of run time and down time compensate for this discrepancy and produce a higher level of performance.

As mentioned in Sect. 1, the OEE percentage of the resources is depending on availability, performance, and quality. The main task in the target system was product distribution and all of the loaded objects into the system were assumed as products

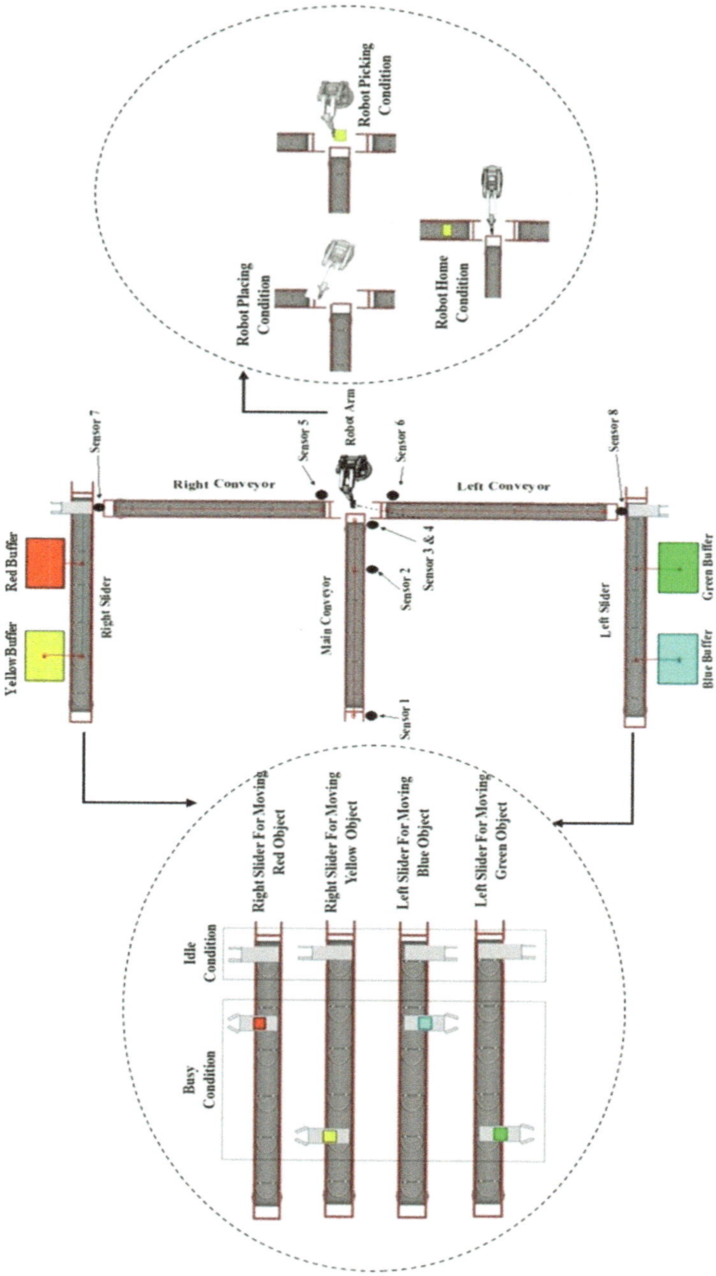

Fig. 24 Arena visualization of the system

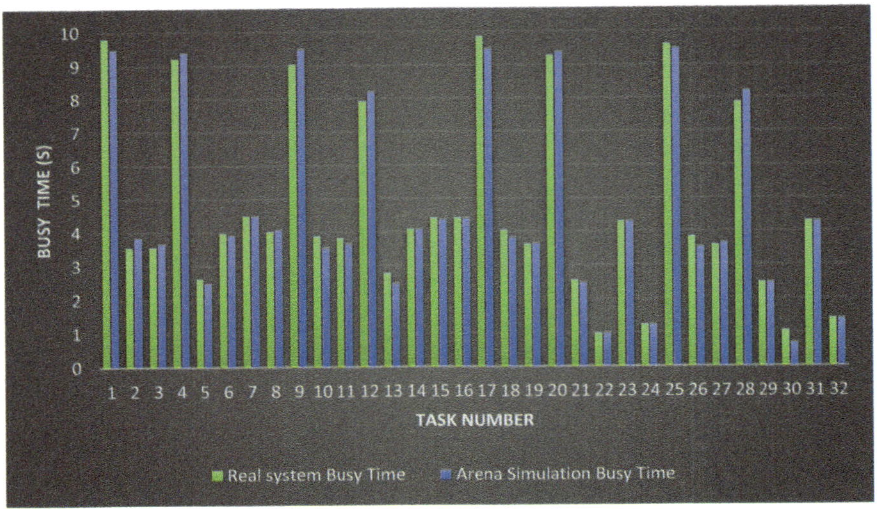

Fig. 25 Time difference between the real system and simulation model busy time

Table 14 Overall and partial system performance based on OEE standard before modification

Recourse	Main conveyor	Left side-conveyor	Right side-conveyor	Robot arm	Right slider for red	Right slider for yellow	Left slider for blue	Left slider for green
Actual time	38.33	15.82	18.52	40.4	12.49	6.55	12.94	6.84
Cycle time	68.79	84.11	68.81	90.95	38.63	52.33	84.11	31.44
Expected time	38.00	15.50	18.50	40.00	12.00	6.00	12.00	6.00
Down time	0.33	0.32	0.52	0.40	0.49	0.55	0.94	0.84
Idle time	30.46	68.29	50.29	50.55	26.14	45.77	71.17	24.6
Planned production time	38.33	15.82	18.52	40.40	12.49	6.56	12.94	6.84
Run time	38.00	15.50	18.00	40.00	12.00	6.01	12.00	6.00
Availability %	99.14	97.98	97.19	99.01	96.08	91.62	92.74	87.72
Ideal cycle time	8.50	7.00	8.50	9.00	11.00	5.00	11.00	5.00
Total count	4.00	2.00	2.00	4.00	1.00	1.00	1.00	1.00
Good count	4.00	2.00	2.00	4.00	1.00	1.00	1.00	1.00
Performance %	89.47	90.32	94.44	90.00	91.67	83.19	91.67	83.33
Quality %	100.00	100.00	100.00	100.00	100.00	100.00	100.00	100.00
OEE %	88.70	88.50	91.79	89.11	88.07	76.22	85.01	73.10

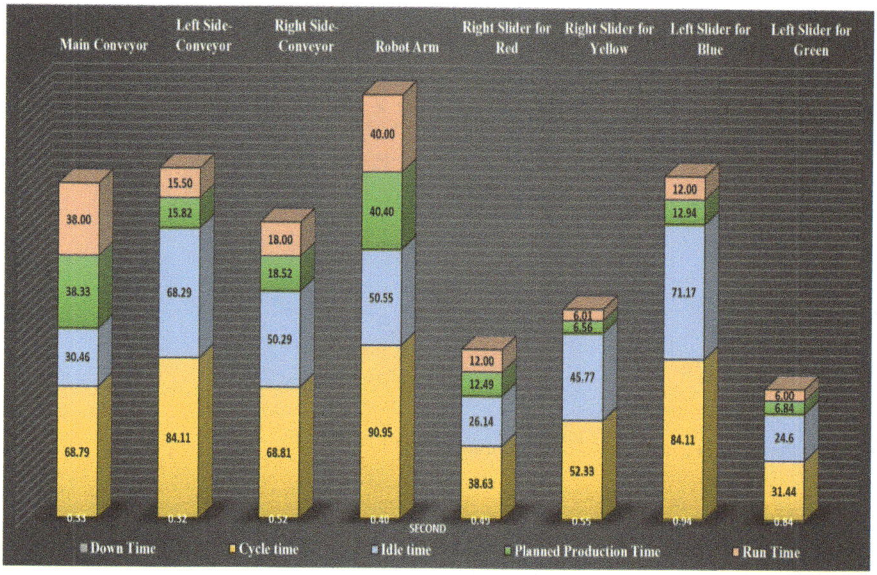

Fig. 26 Overview of resource time for OEE calculation

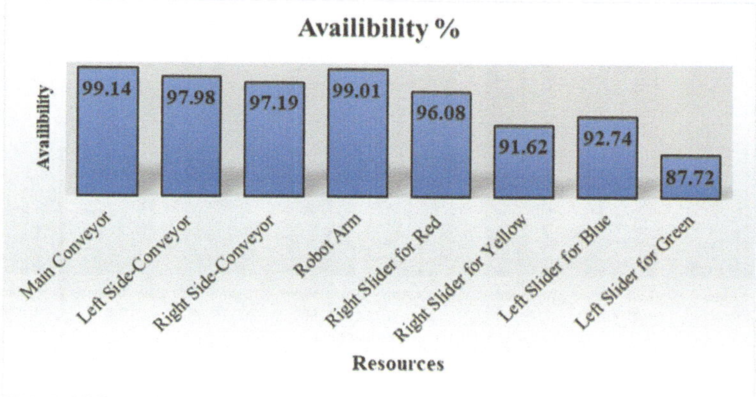

Fig. 27 Availability percentage of the resources

with good quality and the quality percentage for entire resources in the system was considered as 100%. Figure 29 shows the OEE percentage of the resources individually. The effect of the detected problems and limitations in Table 11 is detectable in the OEE percentage for each resource. OEE is affected by availability and performance, and its percentage validates all of the identified problems and limitations. It means for improving and optimizing the system, OEE evaluation is an appropriate standard besides Time study.

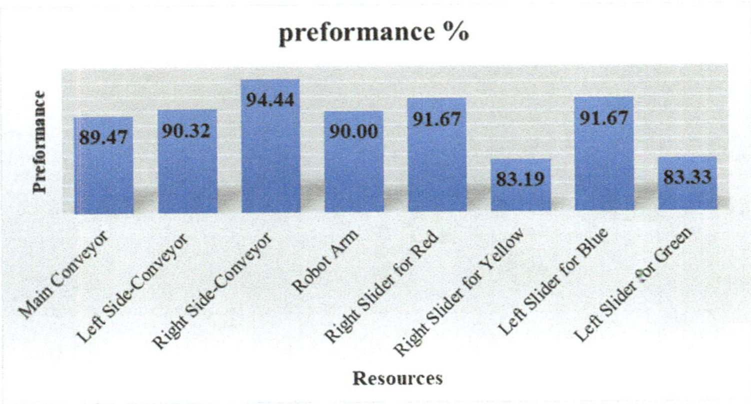

Fig. 28 Performance percentage of the resources

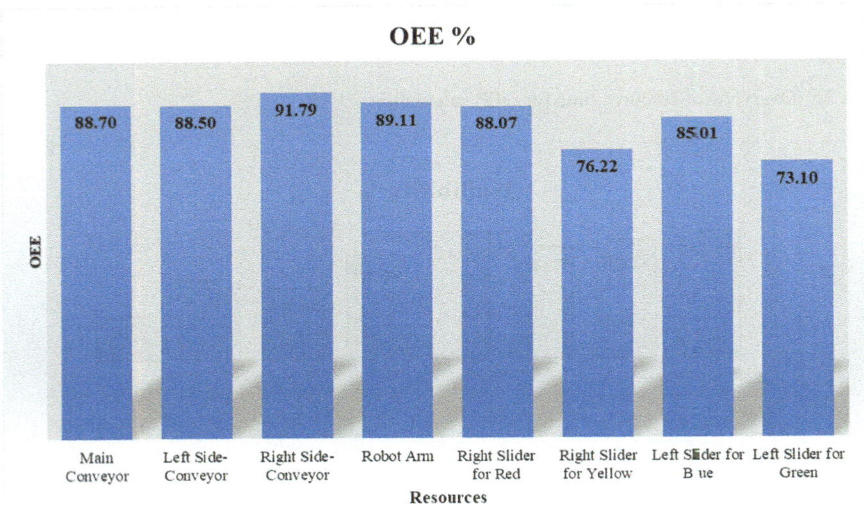

Fig. 29 OEE percentage of the resources

4.6 OEE Analysis After Optimization

Any system modification aimed at resolving the stated flaws and limitations of the target system has a unique effect on the OEE percentage. A function has been built in order to determine the solutions that have the greatest impact on the OEE percentage and system performance. The function contains factors that are directly related to the proposed agent-based architecture. To maximize these factors the master agent or slave agents' software agents must be updated. On the basis of the identified issues and proposed solutions described in Table 11, velocity and distance have been chosen as the target factors. Modifying the distance between resources or the speed of motion necessitates a thorough examination of the system's adaptability to this change. Moreover, if the proposed agent-based control architecture is appropriately adaptable, then the provided solutions may be implemented.

In this study, the distance between resources was reduced to between one-half and two-thirds of the distance in the actual system. Additionally, the speed was increased to its maximum level by taking into account the required modifications for software agents in slaves and interaction protocols in the master agent. Therefore, the factors that can be altered to improve the system's performance are represented as a parameter vector (10).

$$p = [p_1, p_2, p_3] \tag{10}$$

Here, p_2 is the distance between the resources or segments (stations) and p_3 is the speed of motion of the resources. p_1 is the cycle time, including Idle and Busy times (11).

$$T_i, T_b \in p_1 \tag{11}$$

Here, T_{i} is the idle Time and T_b is the busy time. Cycle time is influenced by the velocity and distance between the considered resources; So that to reach to minimum p_1 we have the following:

$$p_1 = \frac{p_2}{p_3} \tag{12}$$

$$p_{1,\min} = \frac{p_{2,\min}}{p_{3,max}} \tag{13}$$

As indicated by the results of the time study of the resources, decreasing the duration of the tasks (Busy Time) will typically result in a reduction of the Idle Times of the system resources. In addition, a resource's idle time is influenced by the performance of its other comparable resources. As the time study and OEE result demonstrated, idle and busy times had a unique effect on the performance of each resource. C_1 is the effectiveness coefficient of idle time, and it demonstrates that idle time has varying effects on different systems. C_2 is regarded as the effectiveness coefficient of busy time, which will be evaluated internally for each resource and will rarely be influenced by non-resource-related elements (14).

$$p_1 = \sum C_1 T_i + C_2 T_b \tag{14}$$

In this study, the offered methods to increase the performance of the system are centred on Time, Velocity, and Distance as effective system factors. These characteristics jointly affect idle and busy time. However, the impact of the changes on a busy time in one resource will be evaluated as a change in idle time in another. To explore the influence of each proposed remedy, C_1 was assumed to be 0 to examine the consequences of concentrating on busy time (15).

$$C_1 = 0 \rightarrow p_1 = \sum C_2 T_b \quad (15)$$

Thus, to reach to minimum $p_{1,}$ p_2 (Distance) should be minimised and p_3 (Velocity) should be maximised.

For the purpose of evaluating the proposed solution, the velocity of the resources has been increased to the highest value allowed by the Arena simulation model. Increasing velocity depends on the capacity and features of the real resource as well as its effect on the preceding and subsequent resources. For instance, the pace of the conveyors was increased to the point where the conveyors could handle the objects precisely, with reduced vibration during transit, and in consideration of the time necessary for the robot to complete the pick-and-place duty. In order to determine the maximum velocity of the conveyor without modifying the system layout, multiple experiments are conducted. By varying the length of conveyors in the simulation model without altering the pace of motion for resources, the impacts of the distance between the resources were explored. Table 15 displays the results of a modified simulation model designed to evaluate the impact of solutions on resource busy times. Figure 30 also depicts a detailed comparison between the time before and after implementing the solutions on the simulation model.

To examine the effect of solutions on OEE percentage, system availability, performance, and quality were necessary. As previously stated, the quality of the objects is deemed to be 100%. And the parameters necessary to calculate the Availability and Performance percentage (Down time, Idle time, Cycle time, Run time, and Planned production time) have been extracted from the simulation model for each solution and resource separately (Table 16).

Figures 31, 32, 33, 34, 35, 36, 37 and 38 are presented to determine the impact of the proposed solution on each resource and select the most effective option. After implementing the solution, this graphical representation aids in analyzing OEE and associated factors. Each graph represents the percentage of Availability, Performance, and OEE, separately.

Analyzing the outcome of the system after modifying the simulation model, determine whether the OEE percentage has been considerably altered by each solution. However, this modification has an immediate effect on system performance and availability. In the vast majority of cases, all options improve the performance of the resource, but not equally.

Changing the distance between the resources (reducing the distances to 2/3 and 1/2 of the actual distance in the simulation model) was anticipated to increase the OEE %. Although these modifications increase the OEE percentages for the majority

Table 15 Task duration on the real system and after modification on the simulation model

Task description	Real	Average of simulation results	2/3 of the real distance change	1/2 of the real distance change	Speed limit change to maximum
Main conveyor handling red object to robot	9.81	9.50	6.63	5.32	8.20
Robot arm picking the red object from main conveyor	3.55	3.84	3.84	3.84	3.25
Robot arm placing red object to right conveyor	3.56	3.66	3.66	3.66	3.20
Right conveyor handling red object to right slider	9.22	9.40	5.60	4.47	8.80
Robot arm moves to its home position after placing red object	2.62	2.50	2.50	2.50	1.99
Main conveyor handling blue object to robot	9.04	9.50	6.63	5.32	8.20
Right slider transfers red object to red buffer	3.98	3.90	2.64	2.03	3.70
Right slider unloading the red object to red buffer	4.50	4.50	4.50	4.50	4.24
Robot arm picking the blue object from main conveyor	3.88	3.55	3.55	3.55	3.34
Robot arm placing blue object to left conveyor	3.82	3.66	3.66	3.66	3.30

(continued)

Table 15 (continued)

Task description	Real	Average of simulation results	2/3 of the real distance change	1/2 of the real distance change	Speed limit change to maximum
Right slider moves to its home position after unloading RED object	4.02	4.08	2.68	2.06	3.90
Left conveyor handling blue object to left slider	7.94	8.33	5.68	4.47	7.50
Robot arm moves to its home position after placing blue object	2.78	2.50	2.50	2.50	1.99
Main conveyor handling yellow object to robot	9.86	9.50	6.63	5.32	8.20
Left slider transfers blue object to blue buffer	4.09	4.09	2.72	2.00	4.02
Left slider unloading the blue object to blue buffer	4.42	4.40	4.40	4.40	4.20
Robot arm picking the yellow object from main conveyor	4.05	3.84	3.84	3.84	3.25
Left slider moves to its home position after unloading blue object	4.43	4.43	2.95	2.27	4.06
Robot arm placing yellow object to right conveyor	3.63	3.66	3.66	3.66	3.20
Right conveyor handling yellow object to right slider	9.30	9.40	5.60	4.47	8.80

(continued)

Table 15 (continued)

Task description	Real	Average of simulation results	2/3 of the real distance change	1/2 of the real distance change	Speed limit change to maximum
Robot arm moves to its home position after placing yellow object	2.57	2.50	2.50	2.50	1.99
Main conveyor handling green object to robot	9.62	9.50	6.63	5.32	8.20
Right slider transfers yellow object to yellow buffer	0.98	0.99	0.62	0.49	0.70
Right slider unloading the yellow object to yellow buffer	4.32	4.32	4.32	4.32	4.10
Robot arm picking the green object from main conveyor	3.84	3.55	3.55	3.55	3.34
Right slider moves to its home position after unloading yellow object	1.25	1.25	0.78	1.20	1.00
Robot arm placing green object to left conveyor	3.60	3.66	3.66	3.66	3.30
Left conveyor handling green object to left slider	7.88	8.33	5.68	4.47	7.50
Robot arm moves to its home position after placing green object	2.50	2.50	2.50	2.50	1.99
Left slider transfers green object to green buffer	1.07	0.71	0.62	0.49	0.66
Left slider unloading the green object to green buffer	4.33	4.32	4.32	4.32	4.10

(continued)

Table 15 (continued)

Task description	Real	Average of simulation results	2/3 of the real distance change	1/2 of the real distance change	Speed limit change to maximum
Left slider moves to its home position after unloading green object	1.44	1.44	1.03	0.82	1.07

of the resources that are connected to a conveyor unit, the OEE percentage decreased for a few of them.

Between sensors 1–2 and 2–3, the primary conveyor contained two segments with varying speeds of motion. Speed modification has the greatest impact on OEE percentage for this resource. Altering the distance between the resources on the primary conveyor without altering the speed boosts OEE as well. These adjustments provide an improvement of 1.8% and 5.28% for reducing the distances to 2/3 and 1/2 of the actual distance, respectively. Increasing the speed of the main conveyor segments has the greatest effect on the OEE percentage, at 8.86%.

Figure 31 demonstrates that the performance percentage of the primary conveyor rose significantly for all proposed alternatives. Nevertheless, the availability of this resource for all revisions dropped marginally. This decrease in availability percentage value is due to its dependence on idle time. On the primary conveyor, idle time is entirely dependent on the percentage of robot availability, which did not change much after implementing the solutions. Therefore, decreasing the distance minimizes the main conveyor's busy time, assuming the idle time does not change owing to robot dependency. In contrast, the performance percentage rose after implementing the solutions since, in all cases, the run time fell while the ideal cycle time remained constant or decreased marginally.

Figure 32 demonstrates that the OEE percentage for the robot arm varies similarly when the offered solutions are implemented on the simulation model. The performance percentage has the greatest influence on the OEE percentage for this resource. The only variable that can affect the robot's OEE percentage is the distance between the point at which the robot picks up objects and the location at which it places them. The enhanced OEE percentages for the robot are 7.39 and 2.98% for reducing the distance to two-thirds and one-half of the real distance, respectively, and 4.70% for altering the robot's motion speed. The improvement in OEE percentage for reducing the distance between pick-and-place points to two-thirds of the actual distance is greater than the improvement for decreasing the distance to half the actual distance. Reducing the distance between these two spots requires a complicated robot motion to correctly reach the parts. Additionally, a complex robot motion requires more time than simple motions.

Figures 33 and 34 demonstrate that the observed variations in the OEE percentage for side conveyors are comparable. Reducing the distance to two-thirds and one-half

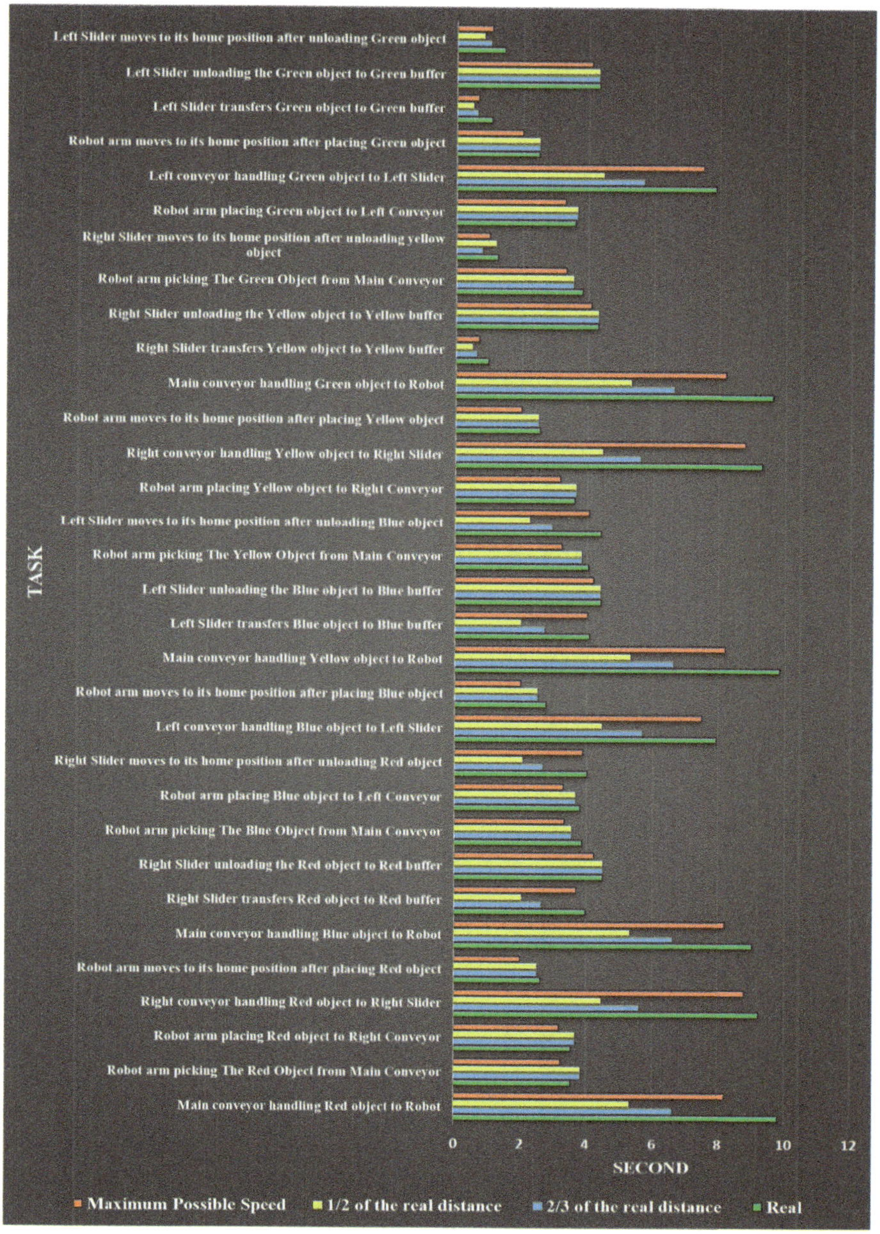

Fig. 30 Comparison of task duration (busy time) before and after the implementation of solutions

Table 16 OEE analysis before and after optimization

Recourse	Cycle time	Down time	Idle time	Planned production time	Run time	Availability (%)	Ideal cycle time	Total count	Good count	Performance (%)	Quality (%)	OEE (%)
Modified to max possible speed												
Main conveyor	58.36	0.33	25.56	32.80	32.47	98.99	8.00	4.00	4.00	98.55	100.00	97.56
Left conveyor	72.50	0.32	57.50	15.00	14.68	97.87	7.00	2.00	2.00	95.37	100.00	93.33
Right conveyor	63.88	0.52	46.28	17.60	17.08	97.05	8.50	2.00	2.00	99.53	100.00	96.59
Robot arm	66.99	0.40	32.80	34.19	33.79	98.83	8.00	4.00	4.00	94.70	100.00	93.59
Right sliders for red	35.29	0.49	23.45	11.84	11.35	95.86	11.00	1.00	1.00	96.92	100.00	92.91
Right sliders for yellow	34.39	0.55	28.59	5.80	5.25	90.52	5.00	1.00	1.00	95.24	100.00	86.21
Left slider for blue	55.07	0.70	42.79	12.28	11.58	94.30	11.50	1.00	1.00	99.31	100.00	93.65
Left slider for green	23.25	0.80	17.42	5.83	5.03	86.28	5.00	1.00	1.00	99.40	100.00	85.76
Modified to 2/3 distance between the resources												
Main conveyor	56.23	0.33	29.71	26.52	26.19	98.76	6.00	4.00	4.00	91.64	100.00	90.50
Left conveyor	69.12	0.32	57.76	11.36	11.04	97.18	5.00	2.00	2.00	90.58	100.00	88.03
Right conveyor	50.90	0.52	39.70	11.20	10.68	95.36	5.00	2.00	2.00	93.63	100.00	89.29
Robot arm	65.94	0.40	26.52	39.42	39.02	98.99	9.50	4.00	4.00	97.39	100.00	96.40

(continued)

Table 16 (continued)

Recourse	Cycle time	Down time	Idle time	Planned production time	Run time	Availability (%)	Ideal cycle time	Total count	Good count	Performance (%)	Quality (%)	OEE (%)
Modified to max possible speed												
Right sliders for red	29.55	0.49	19.73	9.82	9.33	95.01	9.00	1.00	1.00	96.46	100.00	91.65
Right sliders for yellow	27.07	0.55	21.35	5.72	5.17	90.38	5.00	1.00	1.00	96.71	100.00	87.41
Left slider for blue	45.30	0.70	35.23	10.07	9.37	93.05	9.00	1.00	1.00	96.05	100.00	89.37
Left slider for green	29.79	0.80	23.82	5.97	5.17	86.60	5.00	1.00	1.00	96.71	100.00	83.75
Modified to 1/2 distance between the resources												
Main conveyor	50.69	0.33	29.41	21.28	20.95	98.45	5.00	4.00	4.00	95.47	100.00	93.98
Left conveyor	62.37	0.32	53.43	8.94	8.62	96.42	4.00	2.00	2.00	92.81	100.00	89.49
Right conveyor	43.42	0.52	34.55	8.87	8.35	94.14	4.00	2.00	2.00	95.81	100.00	90.19
Robot arm	60.40	0.40	21.28	39.12	38.72	98.98	9.00	4.00	4.00	92.98	100.00	92.02
Right sliders for red	25.88	0.49	17.29	8.59	8.10	94.30	8.00	1.00	1.00	98.77	100.00	93.13
Right sliders for yellow	23.05	0.50	17.54	5.51	5.01	90.93	5.00	1.00	1.00	99.80	100.00	90.74

(continued)

Table 16 (continued)

Recourse	Cycle time	Down time	Idle time	Planned production time	Run time	Availability (%)	Ideal cycle time	Total count	Good count	Performance (%)	Quality (%)	OEE (%)
Modified to max possible speed												
Left slider for blue	39.02	0.65	30.35	8.67	8.02	92.50	8.00	1.00	1.00	99.75	100.00	92.27
Left slider for green	28.98	0.60	23.35	5.63	5.03	89.34	5.00	1.00	1.00	99.40	100.00	88.81
Real system without modification												
Main conveyor	68.79	0.33	30.46	38.33	38.00	99.14	8.50	4.00	4.00	89.47	100.00	88.70
Left conveyor	84.11	0.32	68.29	15.82	15.50	97.98	7.00	2.00	2.00	90.32	100.00	88.50
Right conveyor	68.81	0.52	50.29	18.52	18.00	97.19	8.50	2.00	2.00	94.44	100.00	91.79
Robot arm	78.73	0.40	38.33	40.40	40.00	99.01	9.00	4.00	4.00	90.00	100.00	89.11
Right sliders for red	38.63	0.49	26.14	12.49	12.00	96.08	11.00	1.00	1.00	91.67	100.00	88.07
Right sliders for yellow	36.73	0.55	30.18	6.55	6.00	91.60	5.00	1.00	1.00	83.33	100.00	76.34
Left slider for blue	59.51	0.70	46.57	12.94	12.24	94.59	11.00	1.00	1.00	89.87	100.00	85.01
Left slider for green	31.44	0.80	24.6	6.84	6.04	88.30	5.00	1.00	1.00	82.78	100.00	73.10

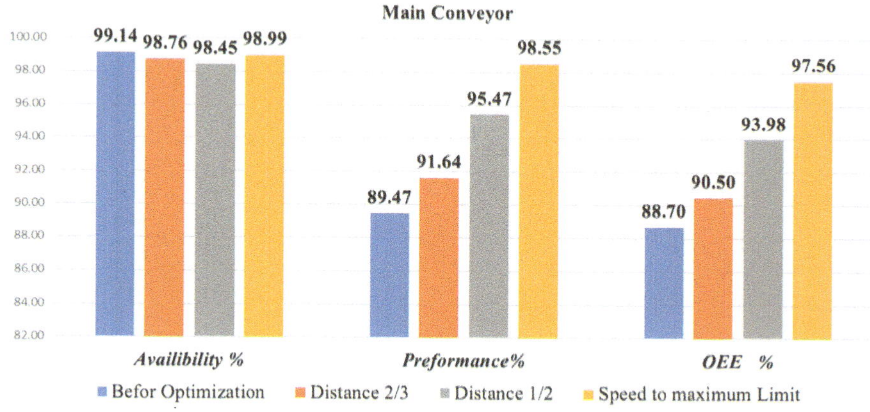

Fig. 31 Comparison of OEE and associated factors for the main conveyor

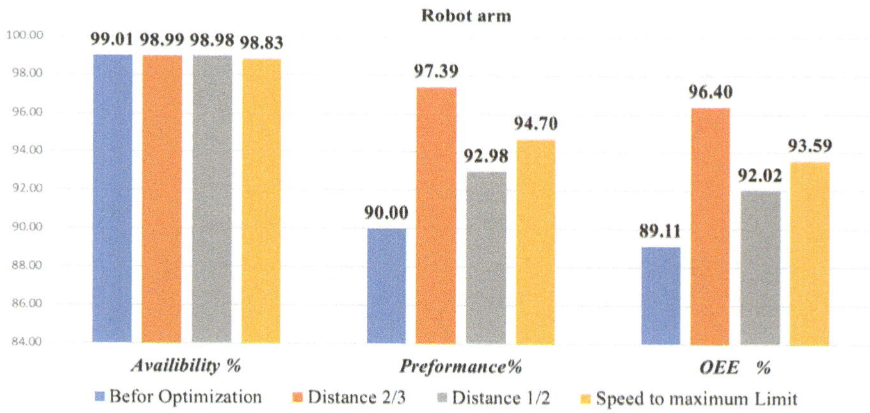

Fig. 32 Comparison of OEE and associated factors for the Robot arm

of the actual distance decreases the OEE percentages by 2.5% and 1.6% on the right side-conveyor, respectively. By altering the maximum speed of this resource, however, the OEE percentage increased by 4.8%. The OEE percentage on the left conveyor practically stayed the same when the distance was altered, but it improved by 4.83% when the speed was increased to maximum.

Figures 33 and 34 demonstrate that adjusting the distances between the two side conveyors decreased the side conveyors' availability. Moreover, the percentage of side-conveyor availability is dependent on robot arm performance. It follows that decreasing the distance between side conveyors reduces the resource's busy time, but the idle time is entirely dependent on robot performance. As shown in Fig. 32, the robot arm's performance percentage improves when the distance is reduced to two-thirds of the real distance. Thus, the availability of side conveyors to reduce the

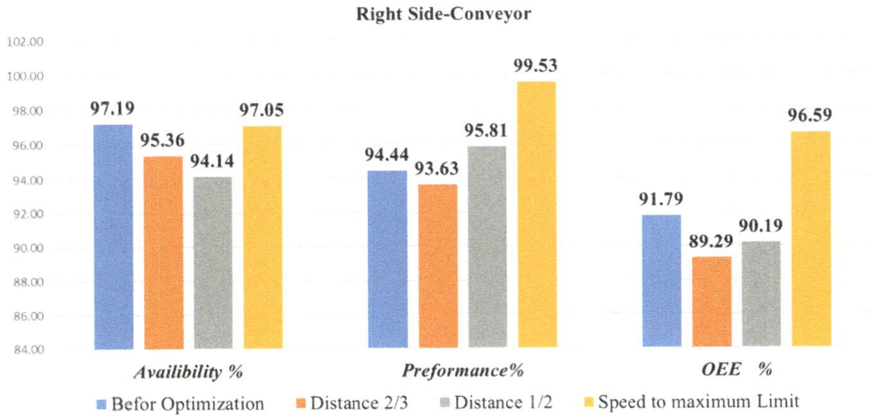

Fig. 33 Comparison of OEE and associated factors for right side-conveyor

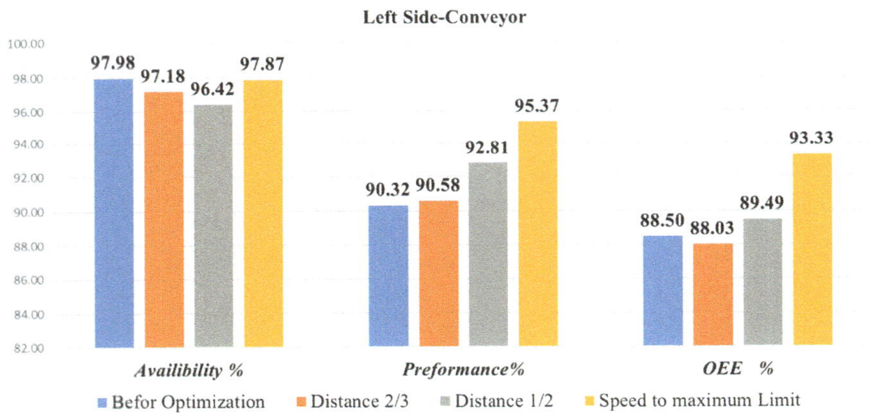

Fig. 34 Comparison of OEE and associated factors for left side-conveyor

distance by 2/3 is greater than 1/2. After changing the distances, the optimal cycle time for side conveyors dramatically decreases, resulting in improved performance.

Regarding the efficacy of a remedy, sliders have wildly divergent attitudes. Changing the spacing between segments enhances the OEE percentage. Right and left sliders respond differently with regard to the effectiveness of methods for moving the portion to buffers near and far from the slider's home position.

Figure 35 demonstrates that the OEE percentage for short range transfer (Red and Blue Objects) increases from 3.58 to 4.43% when the distance is changed to two-thirds and one-half of the actual distance. In contrast, the OEE percentage improvement for the left slider for a limited range of motion ranges between 4.36 and 7.26% when the lengths are reduced to 2/3 and 1/2 of the real distance.

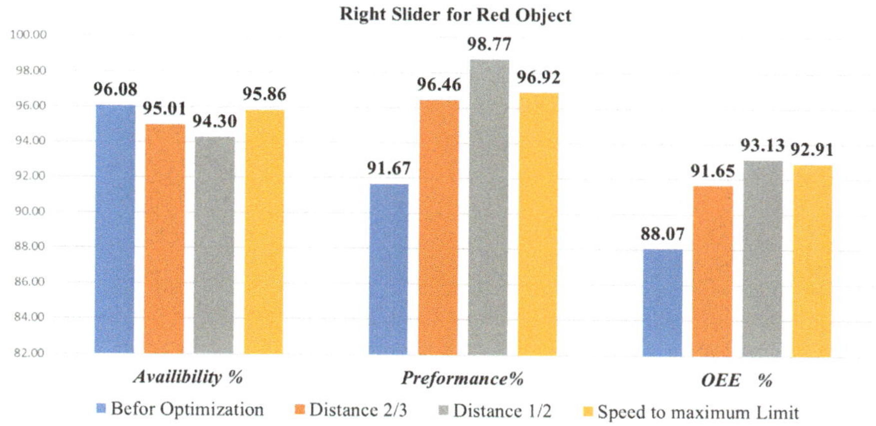

Fig. 35 Comparison of OEE and associated factors for right slider (red object)

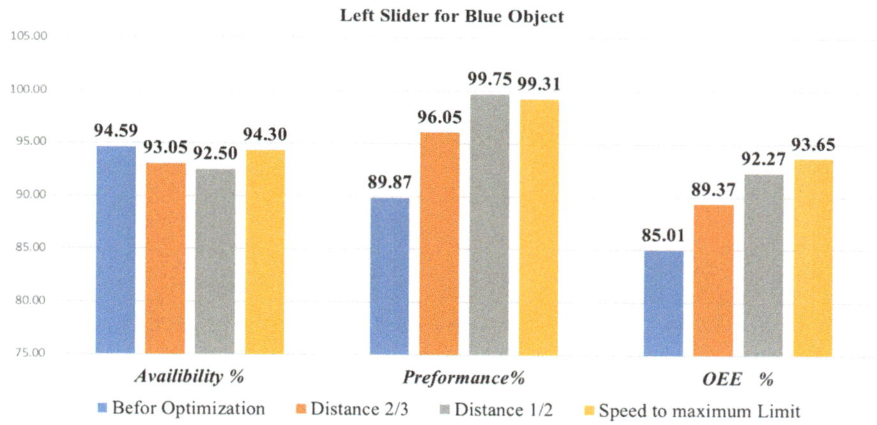

Fig. 36 Comparison of OEE and associated factors for left slider (blue object)

Furthermore, the influence of speed on OEE percentage in the short range of motion for both sliders is significant. Maximum motion speed has a substantial impact on the OEE percentage, which is 4.84% for the right slider and 8.64% for the left slider. Although the availability of sliders for short range motion dropped marginally as a result of their dependence on the performance of the main and side conveyors and robot arm, the performance percentage increased substantially as a result of the reduction in run time for all of the improvements.

Figures 37 and 38 demonstrate that the behaviour of the sliders varies over the long range of motion (Yellow and Green Objects). In contrast to the short range of motion, the OEE percentage of sliders increased more when the distance was altered.

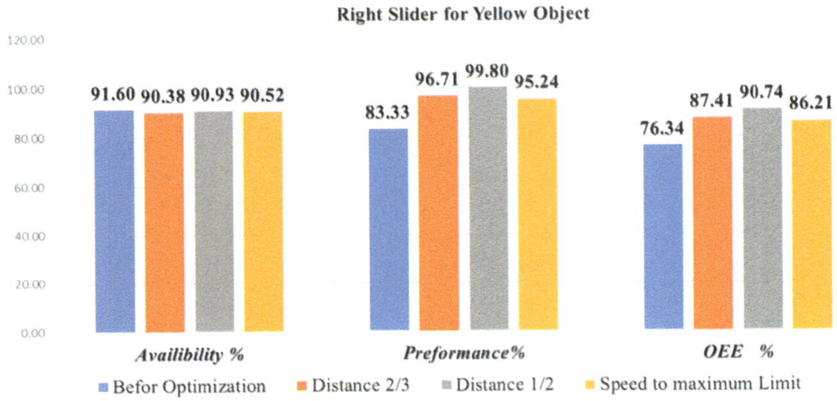

Fig. 37 Comparison of OEE and associated factors for right slider (Yellow object)

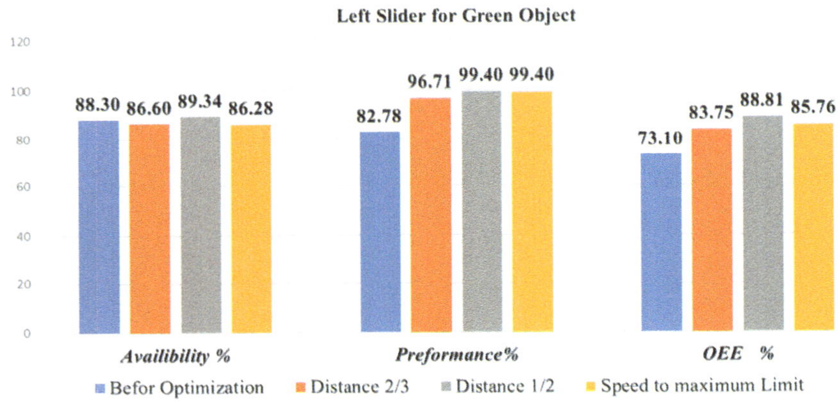

Fig. 38 Comparison of OEE and associated factors for Left slider (Green object)

In addition, the proportion of availability in both sliders over the long range of motion remained relatively unchanged across all adjustments.

Table 17 demonstrates that all recommended solutions have an effect on the OEE percentage. In addition, it is crucial to recognize that the influences on OEE percentages have varying weight and significance for each resource and associated solution. In order to generalize the approach for selecting the optimal optimization factor, a priority has been assigned to each factor. This Priority is determined by the value of OEE percentage improvement. For example, for the main conveyor, we have p_3 (First), $p_{2\text{-}1}$ (Second), and $p_{2\text{-}2}$ (Third), which indicates that the first element with the greatest impact on OEE was speed, and the second and third priorities are reducing the distance by half and two-thirds, respectively.

According to Table 17, increasing the speed of the primary conveyor to its maximum value has the greatest effect on the OEE percentage. However, reducing

Table 17 OEE improvement percentage for each resource

	OEE improvement (%)			Priority of the effectiveness		
Resource	½ of the actual distance (p_{2-1})	$^2/_3$ of the actual distance (p_{2-2})	Speed to max limit (p_3)	First	Second	Third
Main conv	5.28	1.80	8.86	p_3	p_{2-1}	p_{2-2}
Right side-conv	−1.60	−2.50	4.80	p_3	***	***
Left side-conv	0.99	−0.47	4.83	p_3	p_{2-1}	***
Robot arm	2.91	7.29	4.48	p_{2-2}	p_3	p_{2-1}
Right slider (short range)	5.06	3.58	4.21	p_{2-1}	p_3	p_{2-2}
Right slider (long range)	14.40	11.07	9.87	p_{2-1}	p_{2-2}	p_3
Left slider (short range)	7.26	4.36	8.64	p_3	p_{2-1}	p_{2-2}
Left slider (long range)	15.71	10.56	12.66	p_{2-1}	p_3	p_{2-2}

the distance to half its actual size has a greater influence on OEE percentage than reducing the distance to two-thirds of its actual size.

The only factor that can increase the percentage of OEE on a right side-conveyor is motion speed. As seen in Table 17, every change in distance has a negative effect on this resource. Therefore, the only way to increase the OEE percentage for a side conveyor is to increase its speed to its maximum limit. The aforementioned description may be given for the left side-conveyor in light of the fact that reducing the distance to half its real size increases the OEE percentage by a negligible one percent.

Due to the given task and hardware design of the robot arm, reducing the distance between the spots where the robot is picking up and placing objects to two-thirds of their real size enhances the OEE percentage more than any other factor. In addition, raising the robot's speed to its maximum has a greater effect on the OEE percentage than halving the distance.

For slider units, the effect of the target parameters will vary depending on the motion range. According to Table 17, for both ranges of motion in the right slider, reducing the distance to half its actual size has the greatest effect on OEE. However, increasing the speed of motion for short range and reducing the distance to two-thirds of its actual size for long range are of secondary importance for boosting the OEE percentage.

Increasing the speed of motion at short range has the greatest effect on the OEE percentage in the left slider unit. Reducing the distance to half and two-thirds of its real size is the second and third priorities for enhancing OEE, respectively. This behaviour is completely different for the left slider's long range of motion. Reducing the distance to half its actual size has the greatest effect on OEE, while raising the speed and reducing the distance to two-thirds of its actual size have the second and third biggest influence on the OEE percentage, respectively.

Notable is the fact that all recommended methods are implementable if the agent software in slaves and masters and layout design in the target system are updated appropriately. In this research, however, the target system is an example of a SME that is too limited and less flexible to support all of the recommended alternatives. Therefore, it has been advocated that the most effective elements for improving the OEE percentage of the target system be chosen based on the provided priority.

5 Conclusion

This research designs and develops an agent-based control architecture that provides distributed control for a manufacturing system with the traits and characteristics of a small or medium-sized enterprise. The target manufacturing system's distributed control capability demonstrated that the designed control architecture can facilitate the adoption of Industry 4.0 by SMEs.

To adhere to the notion of an agent in the proposed control architecture, a Master–Slave agent-based control architecture was selected. The major concerns in the control architecture were agent task division and communication. The agent-based control architecture using the Master–Slave mechanism consisted of three layers: Physical Resource, Physical resources control, and management.

The objective of the control architecture was to provide sufficient persuasive explanations and solutions to assist SMEs in overcoming their concerns over the modification of their current control system. In order to analyze and enhance the performance of the target system, including its agent-based control architecture, a novel approach was established and deployed.

Successful execution of the technique provided for the identification of all difficulties and limitations related with the integrated target system and agent-based control architecture. In addition, as part of the methodology objectives, the performance of the system was carefully measured, and modifications were made to enhance the system's performance.

The methodology adhered to Lean Six Sigma (DMAIC) standards. As Lean Six Sigma consists of five primary phases, the appropriate techniques and technologies were employed to meet the needs of each phase. For the "Measure" phase of Lean Six Sigma, an accurate combined time study technique was employed to measure the target manufacturing system timing once the agent-based control architecture has been implemented. The results of the time study assisted in identifying the system's flaws and limitations. In addition, the results of the time study were used to assess the performance of the system during the "Analyze" phase of Lean Six Sigma. The challenges and constraints identified by examining the results of the time study revealed the primary methods for overcoming them.

Considering the requirement of the "Analyze" phase of Lean Six Sigma, the Overall Equipment Effectiveness (OEE) standard was used to analyze the system's performance before and after the adoption of the agent-based control architecture. In OEE Standard, the results of the time study are used to calculate the availability and performance of each resource in the production system under consideration. The OEE percentage accurately identified the resources with the most issues and limitations.

After assessing the time study and OEE percentages for each resource and the overall system, as a requirement of the "Improve" phase of Lean Six Sigma, a number of solutions to overcome the identified problem and constraints were determined. Almost every detected issue required a revision of the control architecture to be resolved. The target factor for modifying the control architecture was determined to be the distance between the resources and the speed of their motions.

ARENA was used as a simulation tool to generate a simulation model of the actual system, taking into account the agent-based control architecture. To confirm the simulation model's accuracy, the outcome of the time study was compared to the simulation model's result. All solutions to address the stated difficulties and constraints were tested on a simulation model prior to installation to determine their impact on the system's performance. The modified simulation model's output was utilized by OEE to analyze the effect of the adjustment on the system's performance. Each solution had a unique effect on the performance of the resources. Therefore, a priority was assigned to each solution that indicates the influence of each solution on each resource.

Implementation of the proposed agent-based control architecture and methodology in this research leads to enhanced performance of the target manufacturing system and may enable SMEs to take the first and most significant step toward industry 4.0 adoption.

In this study, the target system was a limited-functionality, small-scale educational manufacturing system. In order to prevent any damage to the system, the established approach required rapid system monitoring and data collecting. In future studies, the proposed methodology can be expanded to real manufacturing systems without the limitations mentioned so that all evaluations can be performed over a longer period of time.

Appendix A

Arena Simulation Control Logic

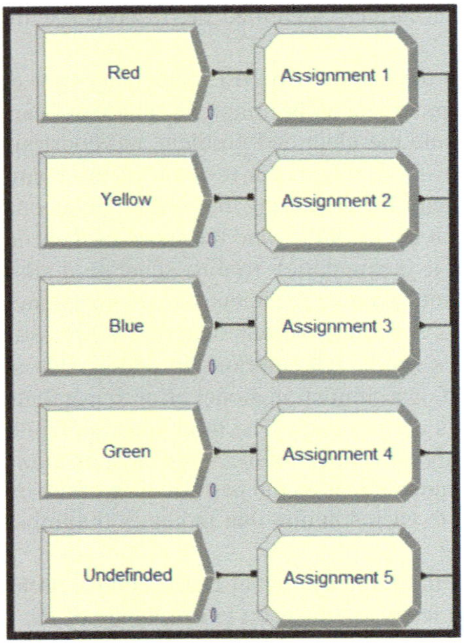

Creating the parts

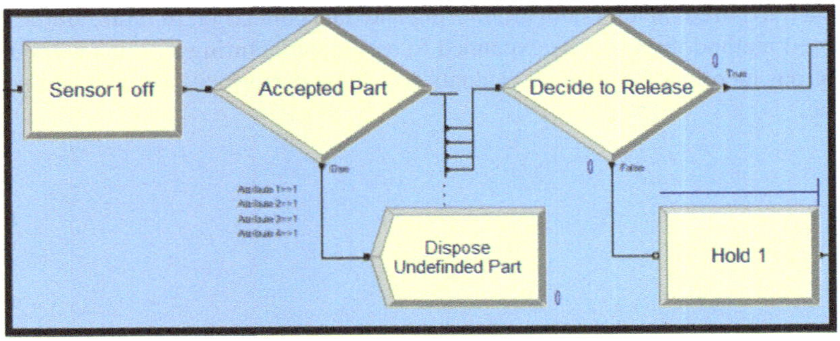

Part loading to main conveyor

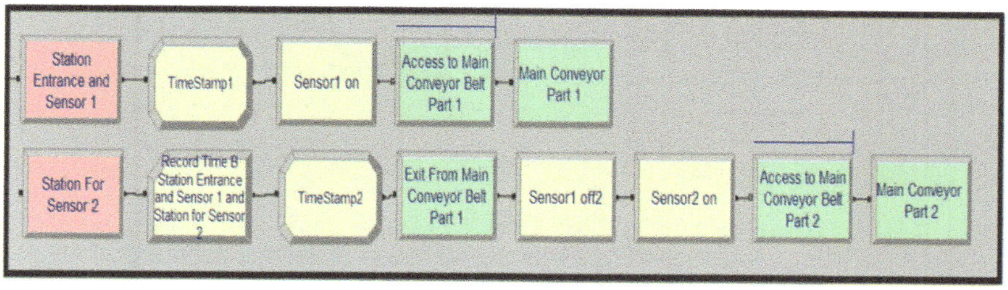

Main conveyor

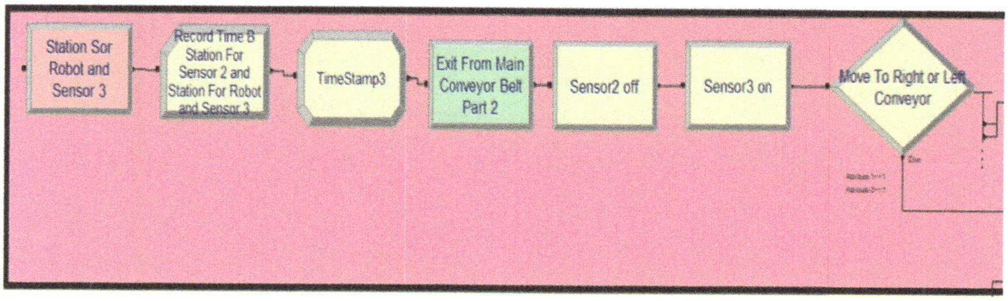

Robot arm (Part 1)

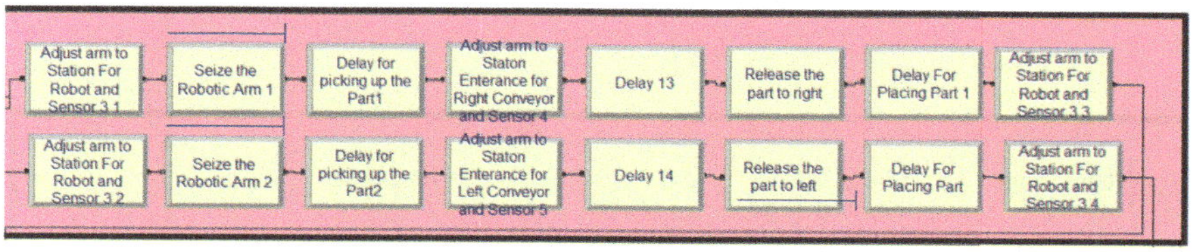

Robot arm (Part 2)

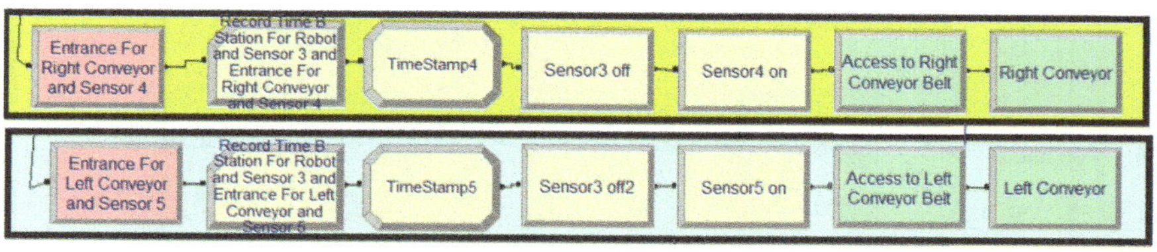

Side conveyors

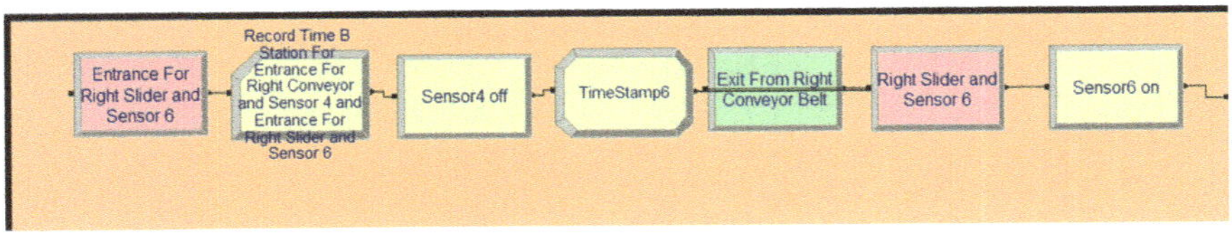

Right slider (Part 1)

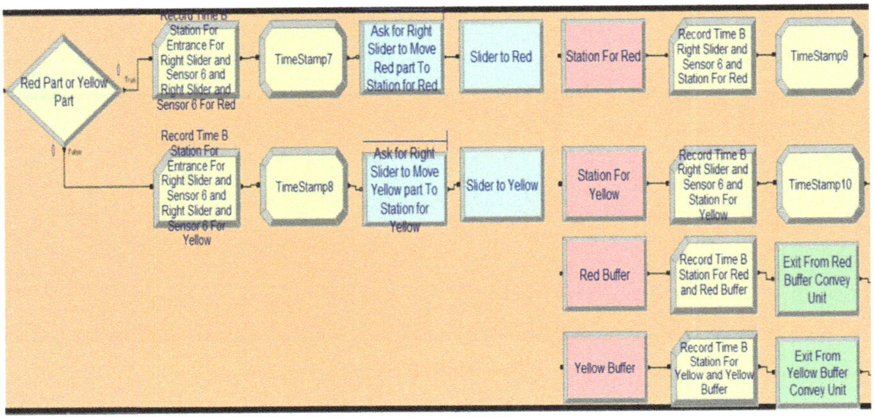

Right slider (Part 2)

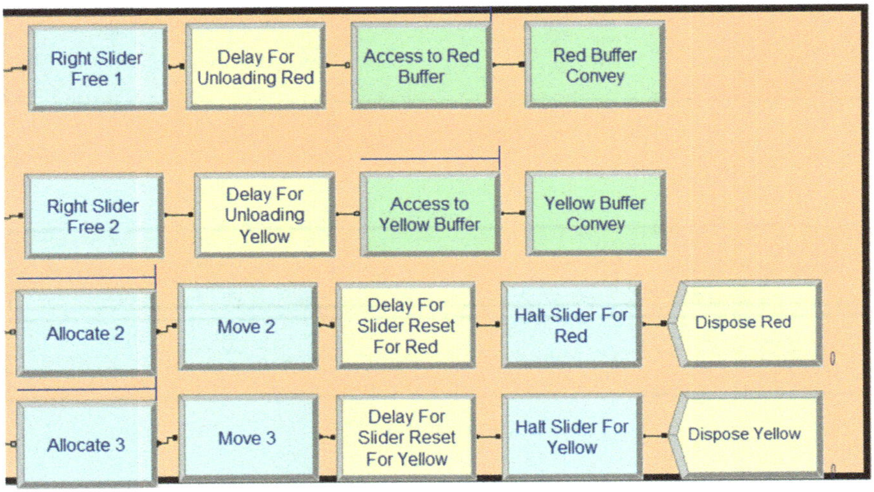

Right slider (Part 3)

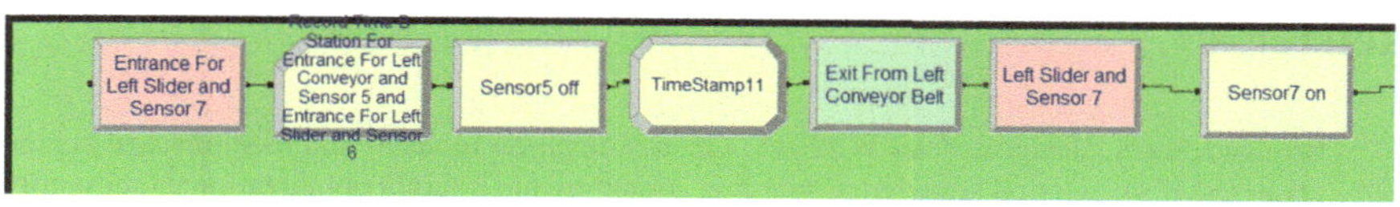

Left slider (Part 1)

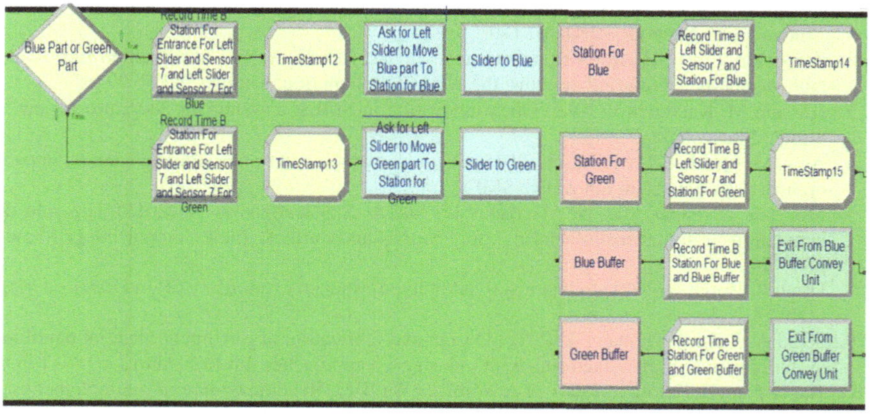

Left slider (Part 2)

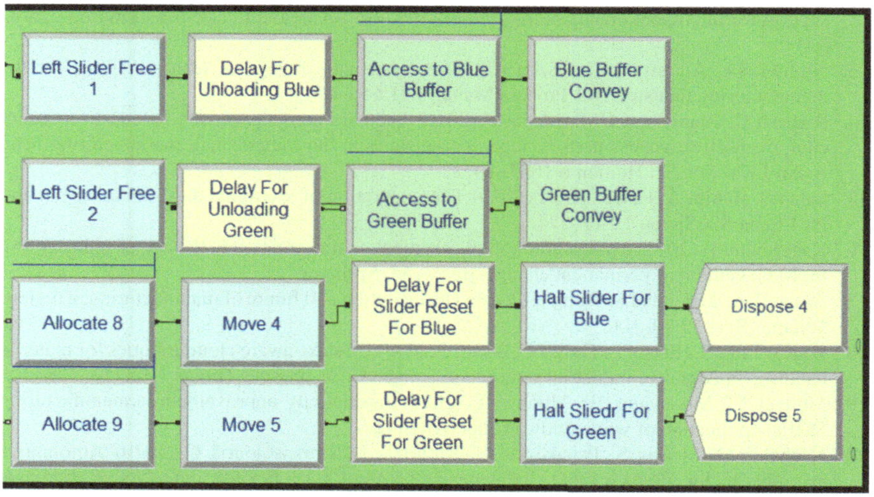

Left slider (Part 3)

References

1. Brunswicker S, Vanhaverbeke W (2015) Open innovation in small and medium-sized enterprises (SMEs): external knowledge sourcing strategies and internal organizational facilitators. J Small Bus Manag 53:1241–1263
2. Jamali D, Lund-Thomsen P, Jeppesen S (2017) SMEs and CSR in developing countries. Bus Soc 56:11–22
3. Azizi A, Yazdi PG, Hashemipour M (2018) Interactive design of storage unit utilizing virtual reality and ergonomic framework for production optimization in manufacturing industry. Int J Interact Des Manufac (IJIDeM) 1–9
4. Barbosa J, Leitão P, Adam E, Trentesaux D (2015) Dynamic self-organization in holonic multi-agent manufacturing systems: the ADACOR evolution. Comput Ind 66:99–111
5. Hossain M, Kauranen I (2016) Open innovation in SMEs: a systematic literature review. J Strateg Manag 9:58–73
6. Ahn JM, Minshall T, Mortara L (2015) Open innovation: a new classification and its impact on firm performance in innovative SMEs
7. Mourtzis D, Papakostas N, Mavrikios D, Makris S, Alexopoulos K (2015) The role of simulation in digital manufacturing: applications and outlook. Int J Comput Integr Manuf 28:3–24
8. Drath R, Horch A (2014) Industrie 4.0: hit or hype? [industry forum]. IEEE Ind Electron Mag 8:56–58
9. Azizi A, Yazdi P, Humairi A (2018) Design and fabrication of intelligent material handling system in modern manufacturing with industry 4.0 approaches. Int Rob Auto J 4:186–195
10. Kagermann H, Helbig J, Hellinger A, Wahlster W (2013) Recommendations for implementing the strategic initiative Industrie 4.0: securing the future of German manufacturing industry; final report of the Industrie 4.0 Working Group, Forschungsunion
11. Hermann M, Pentek T, Otto B (2016) Design principles for industrie 4.0 scenarios. In: 2016 49th Hawaii international conference on system sciences (HICSS), pp 3928–3937
12. Marques M, Agostinho C, Zacharewicz G, Jardim-Gonçalves R (2017) Decentralized decision support for intelligent manufacturing in Industry 4.0. J Ambient Intell Smart Environ 9:299–313
13. Albliwi SA, Antony J, Lim SAH (2015) A systematic review of lean six sigma for the manufacturing industry. Bus Process Manag J 21:665–691
14. Ramesh C, Manickam C, Prasanna S (2016) Lean six sigma approach to improve overall equipment effectiveness performance: a case study in the Indian small manufacturing firm. Asian J Res Soc Sci Human 6:1063–1072
15. Salah S, Rahim A, Carretero JA (2010) The integration of six sigma and lean management. Int J Lean Six Sigma 1:249–274
16. Kagermann H (2015) Change through digitization—value creation in the age of Industry 4.0. In: Management of permanent change, pp 23–45. Springer
17. Esmaeilian B, Behdad S, Wang B (2016) The evolution and future of manufacturing: a review. J Manuf Syst 39:79–100
18. Wan J, Tang S, Hua Q, Li D, Liu C, Lloret J (2018) Context-aware cloud robotics for material handling in cognitive industrial Internet of Things. IEEE Internet Things J 5:2272–2281
19. Munyai TT, Mbonyane BL, Mbohwa C (2017) Productivity improvement in manufacturing SMEs: application of work study. Productivity Press
20. Michalos G, Makris S, Papakostas N, Mourtzis D, Chryssolouris G (2010) Automotive assembly technologies review: challenges and outlook for a flexible and adaptive approach. CIRP J Manuf Sci Technol 2:81–91
21. Islam A, Tedford D, Haemmerle E (2017) Risk determinants of small and mediumsized manufacturing enterprises (SMEs)–an empirical investigation in New Zealand
22. Jazdi N (2014) Cyber physical systems in the context of Industry 4.0. In: 2014 IEEE international conference on automation, quality and testing, robotics, pp 1–4

23. Fields S (1995) A centralized communication system. Nurs Manage 26:48B
24. You X, Hua C, Guan X (2017) Distributed adaptive event-triggered control for leader-following consensus of multi-agent systems. Asian J Control
25. Shen W, Norrie DH, Barthès J-P (2003) Multi-agent systems for concurrent intelligent design and manufacturing. CRC Press
26. Pipattanasomporn M, Feroze H, Rahman S (2009) Multi-agent systems in a distributed smart grid: design and implementation. In: Power systems conference and exposition. PSCE'09. IEEE/PES, pp 1–8
27. Azizi A, Yazdi P, Humairi A (2018) Applications of control engineering in industry 4.0: utilizing internet of things to design an agent based control architecture for smart material handling system. Int Rob Auto J 4:253–257
28. Birukou A, Blanzieri E, Giorgini P (2012) Implicit: a multi-agent recommendation system for web search. Auton Agent Multi-Agent Syst 24:141–174
29. Shen W, Norrie DH (1999) Agent-based systems for intelligent manufacturing: a state-of-the-art survey. Knowl Inf Syst 1:129–156
30. Colman AM (1998) The complexity of cooperation: agent-based models of competition and collaboration. Complexity 3:46–48
31. Andronico A, Carbonaro A, Casadei G, Colazzo L, Molinari A, Ronchetti M (2003) Integrating a multi-agent recommendation system into a mobile learning management system. Proc Artif Intell Mob Syst 123–132
32. Benlghazi A, Chadli E, Moussaid D (2014) Bluetooth technologies for industrial application. In: 2014 5th international conference on information and communication systems (ICICS), pp 1–5
33. Rüßmann M, Lorenz M, Gerbert P, Waldner M, Justus J, Engel P, et al (2015) Industry 4.0: the future of productivity and growth in manufacturing industries. Boston Consulting Group, vol 9, pp 54–89
34. Wang S, Wan J, Li D, Zhang C (2016) Implementing smart factory of industrie 4.0: an outlook. Int J Distrib Sens Netw 12:3159805
35. Geramifard A, Redding J, How JP (2013) Intelligent cooperative control architecture: a framework for performance improvement using safe learning. J Intell Rob Syst 72:83–103
36. Raval SJ, Kant R (2017) Study on lean six sigma frameworks: a critical literature review. Int J Lean Six Sigma 8:275–334
37. Timans W, Ahaus K, van Solingen R, Kumar M, Antony J (2016) Implementation of continuous improvement based on lean six sigma in small-and medium-sized enterprises. Total Qual Manag Bus Excell 27:309–324
38. Antony J, Vinodh S, Gijo E (2016) Lean six sigma for small and medium sized enterprises: a practical guide. CRC Press
39. Taghizadegan S (2010) Essentials of lean six sigma. Elsevier
40. Benjamin SJ, Marathamuthu MS, Murugaiah U (2015) The use of 5-WHYs technique to eliminate OEE's speed loss in a manufacturing firm. J Qual Maint Eng 21:419–435
41. Gibbons PM, Burgess SC (2010) Introducing OEE as a measure of lean six sigma capability. Int J Lean Six Sigma 1:134–156
42. Yaşin MF, Daş GS (2017) KOBİ'lerde ekipman etkinliğinin iyileştirilmesinde TEE tabanlı yeni bir yaklaşım: bir ahşap işleme kuruluşunda uygulama. J Fac Eng Archit Gazi Univ 32:45–52
43. Iannone R, Nenni ME (2013) Managing OEE to optimize factory performance. In: Operations management. InTech
44. Jonsson P, Lesshammar M (1999) Evaluation and improvement of manufacturing performance measurement systems-the role of OEE. Int J Oper Prod Manag 19:55–78
45. Singh M, Narwal M (2017) Measurement of overall equipment effectiveness (OEE) of a manufacturing industry: an effective lean tool. Int J Recent Trends Eng Res 3:268–275
46. Esmaeel RI, Zakuan N, Jamal NM (2018) The mediating role of overall equipment effectiveness on the relationship between fit manufacturing and business performance. Int J Eng Technol 7:1089–1093

47. Ylipää T, Skoogh A, Bokrantz J, Gopalakrishnan M (2017) Identification of maintenance improvement potential using OEE assessment. Int J Product Perform Manag 66:126–143
48. !!! INVALID CITATION !!! .
49. Lesshammar PJM (1999) Evaluation and improvement of manufacturing performance measurement. Int J Oper Prod Manag
50. Demirci T (2012) A model for assessing and evaluating production
51. Demirci T (2012) A model for assessing and evaluating production process effectiveness when applying lean production-a case study
52. Kumar J, Soni V, Agnihotri G (2013) Maintenance performance metrics for manufacturing. Int J Res Eng Technol
53. Lesshammar JA (1999) A model for assessing and evaluating production
54. Silver EA, Pyke DF, Peterson R (1998) Inventory management and production planning and scheduling, vol 3. Wiley, New York
55. Stevenson WJ, Hojati M, Cao J (2007) Operations management, vol 8. McGraw-Hill/Irwin, Boston
56. Syska A (2006) Methods-time-measurement (MTM), Produktionsmanagement: Das A—Z wichtiger Methoden und Konzepte für die Produktion von heute, pp 99–99
57. Zandin KB (2002) MOST work measurement systems. CRC Press
58. Heyde G (1996) Modular arrangement of predetermined time standards. Woodbridge, VA. www.modapts.com
59. Juul-Kristensen B, Hansson G-Å, Fallentin N, Andersen J, Ekdahl C (2001) Assessment of work postures and movements using a video-based observation method and direct technical measurements. Appl Ergon 32:517–524
60. Monden Y (2011) Toyota production system: an integrated approach to just-in-time. CRC Press
61. Negahban A, Smith JS (2014) Simulation for manufacturing system design and operation: literature review and analysis. J Manuf Syst 33:241–261
62. Fui-Hoon Nah F, Lee-Shang Lau J, Kuang J (2001) Critical factors for successful implementation of enterprise systems. Bus Process Manag J 7:285–296
63. Kelton WD, Smith JS, Sturrock DT (2011) Simio & simulation: modeling, analysis, applications. Learning Solutions
64. Fu MC (2002) Optimization for simulation: theory vs. practice. INFORMS J Comput 14:192–215
65. Juan AA, Faulin J, Grasman SE, Rabe M, Figueira G (2015) A review of simheuristics: extending metaheuristics to deal with stochastic combinatorial optimization problems. Oper Res Perspect 2:62–72
66. Xu J, Huang E, Hsieh L, Lee LH, Jia Q-S, Chen C-H (2016) Simulation optimization in the era of Industrial 4.0 and the industrial internet. J Simul 10:310–320
67. Brettel M, Friederichsen N, Keller M, Rosenberg M (2014) How virtualization, decentralization and network building change the manufacturing landscape: an Industry 4.0 perspective. Int J Mech Ind Sci Eng 8:37–44
68. Schriber TJ, Brunner DT, Smith JS (2014) Inside discrete-event simulation software: how it works and why it matters. In: Proceedings of the 2014 winter simulation conference, pp 132–146
69. Alfieri A, Matta A, Pedrielli G (2015) Mathematical programming models for joint simulation–optimization applied to closed queueing networks. Ann Oper Res 231:105–127
70. Alfieri A, Matta A (2012) A time-based decomposition algorithm for fast simulation with mathematical programming models. In: Proceedings of the winter simulation conference, p 231
71. Figueira G, Almada-Lobo B (2014) Hybrid simulation–optimization methods: a taxonomy and discussion. Simul Model Pract Theory 46:118–134
72. Wang S, Wan J, Zhang D, Li D, Zhang C (2016) Towards smart factory for industry 4.0: a self-organized multi-agent system with big data based feedback and coordination. Comput Netw 101:158–168

73. Prinz C, Morlock F, Freith S, Kreggenfeld N, Kreimeier D, Kuhlenkötter B (2016) Learning factory modules for smart factories in industrie 4.0. Proc CIRP 54:113–118
74. Santos C, Mehrsai A, Barros A, Araújo M, Ares E (2017) Towards Industry 4.0: an overview of European strategic roadmaps. Proc Manuf 13:972–979
75. Moeuf A, Pellerin R, Lamouri S, Tamayo-Giraldo S, Barbaray R (2018) The industrial management of SMEs in the era of Industry 4.0. Int J Prod Res 56:1118–1136
76. Bechtsis D, Tsolakis N, Vouzas M, Vlachos D (2017) Industry 4.0: sustainable material handling processes in industrial environments. In: Computer aided chemical engineering, vol 40, pp 2281–2286. Elsevier
77. Chan F, Ip R, Lau H (2001) Integration of expert system with analytic hierarchy process for the design of material handling equipment selection system. J Mater Process Technol 116:137–145
78. Lewandowski M, Gath M, Werthmann D, Lawo M (2013) Agent-based control for material handling systems in in-house logistics-towards cyber-physical systems in in-house-logistics utilizing real size. In: Smart SysTech 2013; European conference on smart objects, systems and technologies, pp 1–5
79. Johnstone M, Creighton D, Nahavandi S (2010) Status-based routing in baggage handling systems: searching verses learning. IEEE Trans Syst Man Cybern Part C (Appl Rev), 40:189–200
80. Lau HY, Woo S (2008) An agent-based dynamic routing strategy for automated material handling systems. Int J Comput Integr Manuf 21:269–288
81. Jain S, Triantis KP, Liu S (2011) Manufacturing performance measurement and target setting: a data envelopment analysis approach. Eur J Oper Res 214:616–626
82. Wakjira MW, Singh AP (2012) Total productive maintenance: a case study in manufacturing industry. Glob J Res Eng 12
83. Hon K (2005) Performance and evaluation of manufacturing systems. CIRP Ann 54:139–154
84. Abdi MR, Labib AW (2011) Performance evaluation of reconfigurable manufacturing systems via holonic architecture and the analytic network process. Int J Prod Res 49:1319–1335
85. Bol JC, Kramer S, Maas VS (2016) How control system design affects performance evaluation compression: the role of information accuracy and outcome transparency. Acc Organ Soc 51:64–73
86. Swarnakar V, Vinodh S (2016) Deploying lean six sigma framework in an automotive component manufacturing organization. Int J Lean Six Sigma 7:267–293
87. Habidin NF, Mohd Yusof SR, Mohd Fuzi N (2016) Lean six sigma, strategic control systems, and organizational performance for automotive suppliers. Int J Lean Six Sigma 7:110–135
88. Franchetti MJ (2015) Lean six sigma for engineers and managers: with applied case studies. CRC Press
89. Lande M, Shrivastava R, Seth D (2016) Critical success factors for lean six sigma in SMEs (small and medium enterprises). TQM J 28:613–635
90. Al-Saleh KS (2011) Productivity improvement of a motor vehicle inspection station using motion and time study techniques. J King Saud Univ Eng Sci 23:33–41
91. Gružauskas V, Karosevičiūtė D, Srovnalíková P (2016) Labour and machine efficient utilization importance to the enterprise profit. J Manag 28
92. Elnekave M, Gilad I (2006) Rapid video-based analysis system for advanced work measurement. Int J Prod Res 44:271–290
93. Bon AT, Daim D (2010) Time motion study in determination of time standard in manpower process
94. Longo F, Mirabelli G (2009) Effective design of an assembly line using modelling and simulation. J Simul 3:50–60
95. Kuhlang P, Edtmayr T, Sihn W (2011) Methodical approach to increase productivity and reduce lead time in assembly and production-logistic processes. CIRP J Manuf Sci Technol 4:24–32
96. de Almeida DL, Ferreira JC (2009) Analysis of the methods time measurement (MTM) methodology through its application in manufacturing companies. In: 19th international conference on flexible automation and intelligent manufacturing (FAIM). Middlesbrough

97. Karim AM, Tuan ST, Emrul Kays H (2016) Assembly line productivity improvement as re-engineered by MOST. Int J Prod Perform Manage 65:977–994
98. Gupta MPK, Chandrawat MSS (2012) To improve work force productivity in a medium size manufacturing enterprise by MOST technique. IOSR J Eng 2:8–15
99. Gibbons PM (2006) Improving overall equipment efficiency using a lean six sigma approach. Int J Six Sigma Compet Adv 2:207–232
100. Mandahawi N, Fouad RH, Obeidat S (2012) An application of customized lean six sigma to enhance productivity at a paper manufacturing company. JJMIE 6:103–109
101. Ghafoorpoor Yazdi P, Azizi A, Hashemipour M (2018) An empirical investigation of the relationship between overall equipment efficiency (OEE) and manufacturing sustainability in industry 4.0 with time study approach. Sustainability 10
102. Puvanasvaran A, Mei C, Alagendran V (2013) Overall equipment efficiency improvement using time study in an aerospace industry. Proc Eng 68:271–277
103. Patel N (2015) Reduction in product cycle time in bearing manufacturing company. Int J Eng Res Gen Sci 3:466–471
104. Yang K, Basem S, El-Haik B (2003) Design for six sigma. McGraw-Hill, New York
105. Naeem K, Ullah M, Tariq A, Maqsood S, Akhtar R, Nawaz R et al (2016) Optimization of steel bar manufacturing process using six sigma. Chin J Mech Eng 29:332–341
106. Jayant A, Kumar S (2012) Simulation modeling of integrated supply chain logistics networks. In: Proceedings of the international conference on soft computing for problem solving (SocProS 2011), December 20–22, 2011, pp 101–111
107. Zahraee SM, Shariatmadari S, Ahmadi HB, Hakimi S, Shahpanah A (2014) Application of design of experiment and computer simulation to improve the color industry productivity: case study. J Teknol 68
108. Maropoulos PG, Ceglarek D (2010) Design verification and validation in product lifecycle. CIRP Ann 59:740–759
109. Sargent RG (2013) Verification and validation of simulation models. J Simul 7:12–24
110. Macal CM (2016) Everything you need to know about agent-based modelling and simulation. J Simul 10:144–156

Path Tracking Control of Bioflexible Probes Exposed to Uncertainties and Internal Tissues Disturbances with Unknown Upper Bonds Using Robust-Adaptive Sliding Mode Control

Zahra Arabtelgerd, Abbasali Koochakzadeh, Mojtaba Naderi Soorki, and Seyed Mohammad Yasoubi

Abstract Flexible probes, for the purpose of minimally invasive surgery by entering the skin with minimal damage to internal tissues, have attracted the attention of researchers. One of the challenges of such bioinspired probes is the existence of model uncertainties and disturbances such as the unknown applied force of internal tissues. In this paper, the optimal trajectory tracking problem among these probes when subject to both model uncertainties and unbounded external disturbances is considered. For this purpose, assuming that the upper band of the scalar sum of uncertainties and disturbances is unknown, the adaptive robust sliding mode control rules have been designed to track the position and prove the stability in such a way that this upper band is estimated using adaptive rules. Furthermore, in order to make the problem more realistic and considering the dependence of many external disturbances on the state of the probe inside the body, the upper band of system disturbances is not a fixed value, but a linear function with unknown coefficients of the soft variables of the probe state has been taken and the robust-adaptive sliding mode control law has been designed for stabilization and estimation of functional upper band coefficients. The results of numerical simulations show the correctness of the designed controllers.

Z. Arabtelgerd
Department of Aerospace Engineering, K. N. Toosi University of Technology, Tehran, Iran
e-mail: zahra.arabtelgerd@email.kntu.ac.ir

A. Koochakzadeh
Department of Electrical Engineering, Amirkabir University of Technology (Tehran Polytechnic), Tehran, Iran
e-mail: Abbaskz@aut.ac.ir

M. N. Soorki (✉)
Department of Electrical Engineering, Sharif University of Technology, Tehran, Iran
e-mail: mojtabanaderi@aut.ac.ir

S. M. Yasoubi
Department of Mechanical Engineering, Mashhad Branch Islami Azad University of Mashhad, Mashhad, Iran
e-mail: Mohammad.yasoubi@mashdiau.ac.ir

© The Author(s), under exclusive license to Springer Nature Singapore Pte Ltd. 2023

A. Azizi (ed.), *Control Engineering in Mechatronics*, Emerging Trends in Mechatronics,
https://doi.org/10.1007/978-981-16-7775-5_2

Keywords Flexible probes · External disturbances · Model uncertainties · Optimal path tracking · Robust-adaptive control · Sliding mode control

1 Introduction

Inspired by nature, bio-inspired flexible probes are used to inject and sample tissues in contact with the patient's skin [1]. Flexible guided probes solve many problems that surgeons face during the operation due to the use of conventional rigid probes and needles. In addition, their use leads to achieving the goal, new treatment methods and reducing the rate of complications [2].

In the bioinspired flexible probe, the direction of the probe is controlled by its steerable tip. With this approach, the radius of curvature of the probe movement is changed by using the relative position between different parts of the probe, which is called command offset [3, 4]. In the literature, various methods for controlling such probes have been presented by researchers. Recently, in [5], the fuzzy control method was used for the sequence on the desired two-dimensional path for the controllable flexible needle probe, and also in [6] the adaptive control law is applied to bio-inspired flexible probe tip tracking in the presence of model uncertainties. In [7], the control of bio-inspired probes with robust control in the presence of disturbances is discussed. However in this article, knowledge of the upper band of disturbances is required for the control law and the state variables are also controlled independently, which will cause problems in practice.

The study and control of flexible needles and probes have been of great interest of many control researchers in the last two years. In [8], a three-dimensional model of a probe with a diagonal tip has been simulated to check its performance in relation to the tissues of the simulation model and experimentally validated by testing in soft gel.

In [9], the idea of back-stepping control for a 3D steering needle is presented. In [10], a particle swarm optimization method is used for flexible needle routing. In this reference, in fact, the arcuate radius of the flexible probe path has been calculated and the particle optimization method has been used to plan the path by changing the rotation angle of the needle.

Finally, a three-dimensional feedback control using fuzzy logic for a flexible needle with a controllable diagonal tip is designed in [11]. In the last one or two years, the robust sliding mode control has received more attention from researchers for the control of flexible probes and needles. Among these, Refs. [12–14] can be mentioned. In [12, 13], taking into account the kinematic model of flexible needles in contact with the tissue, the hyper-twisting sliding mode control method is used, which is an idea to reduce chattering, and its results in reducing chattering are compared with normal sliding mode control. The same authors in [14], by adding the sliding mode observer to the hyper-torsional sliding mode control to deal with external disturbances, have dealt with the control of flexible needles with diagonal tips and the simulation and practical implementation of the method.

This literature survey shows that the assumption of the existence of both uncertainties and external disturbances in the model under stimulation of the biologically flexible feather; in which the upper band of disturbances and uncertainties are unknown and especially—considering that the applied forces from other tissues in the path of the probe act similarly to the disturbance and are dependent on the location of the probe—that the upper band of disturbances is not a scalar value but a function of soft modes with unknown coefficients is still one of the open issues in this field, which will be discussed in this article.

A usual method to deal with the dynamic systems that are exposed to modeling uncertainties and external disturbances with unknown upper bounds, is sliding ode control and robust adaptive sliding mode control [15–17]. In [18] fast finite time adaptive sliding mode control of a quad rotor in the presence of uncertainties and disturbances is considered. It is assumed that the upper bound of the model uncertainties and external disturbances is a nonlinear function with unknown coefficients. These coefficients are estimated via stable adaptive laws. The authors in [16] had designed a new robust-adaptive fractional order sliding mode control in order to achieve consensus in a fractional order swarm system in the presence of modeling uncertainties and external disturbances. The upper bound of perturbation is considered to be unknown in this reference. Also in [17] asymptotic stabilization of uncertain nonlinear fractional-order systems with bounded inputs in the presence of model uncertainties and external disturbances is addressed. The main contribution of this paper is to develop a new bounded fractional-order chattering free adaptive sliding mode control in which the system states and upper bound of disturbances is estimated using adaptive laws.

In the last two or three years, many researchers have studied the flexible probes themselves, maybe under other titles such as flexible needle and among which we can refer to the papers [19–21] that study the structure and control process of this type of flexible probes. Nevertheless, there are not many references about their robust or adaptive control in the presence of disturbances and uncertainties. In this context, Ref. [22] may be mentioned as one of the few papers in which the method of the super-twisting sliding mode robust control law based on a sliding mode observer is presented for path tracking of flexible probes.

Therefore, the control of these probes in the presence of perturbations (meaning disturbances and uncertainties that are caused by unknown forces of fat tissue, veins and muscles in the probe path inside the body) with unknown upper bond and in the conditions where the upperbond of this uncertaintes is not only a scaler value but a function of system states with unknown coefficients, is one of the open problems in the field of control of these probes.

In this article, the aim is to design a new adaptive sliding mode control rule, in which, along with the probe control for convergence to the desired path, deals with uncertainties and disturbances with the upper band of an unknown function, and the coefficients of this upper band with the rules will be estimated accordingly.

In general, the innovation of this research can be considered as the following main items:

(1) *Bio-inspired flexible probe control with the under excitation equations in following the desired path considering both uncertainty and disturbance.*
(2) *Dealing with uncertainty and disturbance with an unknown upper band in the probe model and estimating this upper band with adaptive rules.*
(3) *Considering the upper band of system disturbances as a function of system states with unknown coefficients and estimating the coefficients of this function with adaptive rules.*
(4) *Designing a new robust adaptive sliding mode control law for tracking the desired path in biological probe.*

This paper is organized as follows: in Sect. 2, problem formulation with probe modeling in the presence of uncertainty and disturbance is presented. In Sect. 3, the design of robust adaptive sliding mode control is done. The simulation results are shown in Sect. 4 and finally in Sect. 5 conclusion will be given.

2 Problem Formulation

The kinematic model of a bioinspired flexible probe is expressed using the following nonlinear equations [23, 24]:

$$\begin{bmatrix} \dot{x} \\ \dot{y} \\ \dot{\theta} \\ \dot{\delta} \end{bmatrix} = \begin{bmatrix} \cos\theta \\ \sin\theta \\ k(\delta - \varepsilon\theta) \\ 0 \end{bmatrix} v_1 + \begin{bmatrix} 0 \\ 0 \\ 0 \\ 1 \end{bmatrix} v_2 \tag{1}$$

where θ is the angle of the tip of the probe, δ expresses the steering offset, constant coefficient $k(\text{mm}^{-2})$ expresses the coefficient between the steering offset and the curvature of the probe, ε (mm) is the distance between the axis of the two parts and expressing the compensation coefficient caused by the shape of the probe in this model, and v_1 and v_2 show the forward linear speed of the probe and the rate of change of the steering offset, respectively (Fig. 1).

In controlling such probes, the goal is that the surgeon or the operator controls the probe in such a way that its tip reaches a desired place inside the tissue and the injection operation is performed there. As a result, the tip of the probe, which is the point of effect, must follow a predetermined optimal path to reach the target. It is possible to control and guide the angle of its tip by changing the offset between different parts of the probe. That is, by changing the steering offset at the end of the probe, the angle of the tip of the probe changes and it is possible to guide it. With the definition as the generalized coordinates and as the input velocities of the system, model (1) can be written as follows [23, 24]:

$$\dot{q} = G(q)u \tag{2}$$

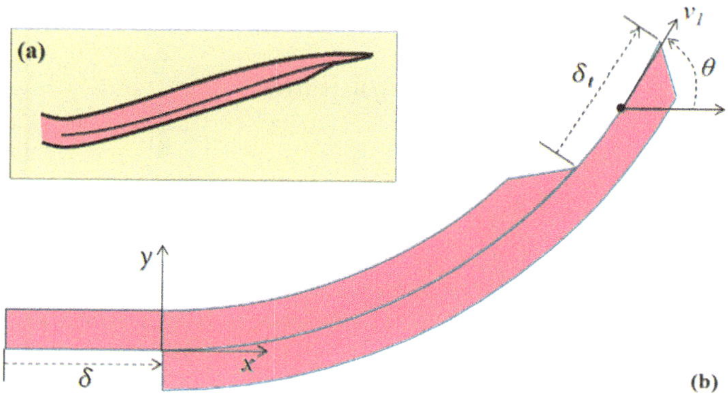

Fig. 1 Bioinspired flexible probe. **a** Probe following the desired path, **b** Flexible probe model

where

$$G(q) = \begin{bmatrix} \cos\theta & 0 \\ \sin\theta & 0 \\ \mathrm{k}(\delta - \varepsilon\theta) & 0 \\ 0 & 1 \end{bmatrix} \tag{3}$$

In addition to the lack of accurate knowledge of the parameters of model (2) in practice, this model may not accurately express the behavior of the probe. For example, the relationship between the instantaneous curvature (ρ) and the command offset (δ_t) at the tip of the probe is assumed as a linear function with a constant coefficient $k(\text{mm}^{-2})$. In practice this relationship is more complicated and with this assumption, a part of the motion behavior of the probe has been ignored. These cases lead us to the existence of model uncertainties in the probe model.

In addition to the uncertainty, due to the movement of the probe tip inside the tissue and the lack of accurate information about the details of the tissue in which the probe moves, model (2) may be subject to noise and disturbances. This disturbance may be the presence of vessels and unpredictable moving tissues in the direction of the probe movement, which has affected its movement, and on the other hand, its nature is unclear.

Therefore, taking into account modeling uncertainties and external disturbances, model (2) can be written as follows:

$$\dot{q} = (G(q) + \Delta G(q))u + E(t) \tag{4}$$

In this equation, $\Delta G(q) \in R^{4\times 2}$ is the unmodeled uncertainty and kinematics, and $E(t) \in R^4$ indicates the disturbances or noises in the probe model, which have no information about the nature and their model are not available and it is only assumed that these disturbances are limited. By collecting uncertainties and disturbances, the final model of the system in (4) can be considered as follows:

$$\dot{q} = G(q)u + d(t) \tag{5}$$

which we have in this model:

$$d(t) = \Delta G(q)u + E(t) \tag{6}$$

$d(t) = [d_1(t), d_2(t), d_3(t), d_4(t)]^T$ expresses the sum of model uncertainties and disturbance and noise which are applied to the system. Now, consider the following assumption:

Assumption 1 Assume that the amount of external disturbances and uncertainties is limited as follows:

$$|d_i(t)| \le \varepsilon_i \qquad i = 1, \ldots, 4 \tag{7}$$

where $\varepsilon_i \in R^+$ represents the upper band of the set of uncertainty and disturbance and unknown. Now according to (2) and (7), the probe model exposed to disturbances for each of the state variables can be writt en as follows:

$$\dot{x} = \cos(\theta)u_1 + d_1, \qquad |d_1(t)| \le \varepsilon_1 \tag{8}$$

$$\dot{y} = \sin(\theta)u_1 + d_2, \qquad |d_2(t)| \le \varepsilon_2 \tag{9}$$

$$\dot{\theta} = k(\delta - \varepsilon\theta)u_1 + d_3 \qquad |d_3(t)| \le \varepsilon_3 \tag{10}$$

$$\dot{\delta} = u_2 + d_4 \qquad |d_4(t)| \le \varepsilon_4 \tag{11}$$

The goal is to design the control law of the sliding model in the probe model under uncertainty and disturbance in (8)–(11) in such a way that despite Assumption 1, the location of the tip of the probe i.e., $[x, y]^T$ follows the desired direction $[x_{des}, y_{des}]^T$.

3 Design of Robust-Adaptive Sliding Mode Controller

3.1 Fixed Upper Band and Unknown Disturbances

In this section, considering Assumption 1, the design of the resistant-adaptive sliding mode control law for the probe will be discussed, in which the upper band of the sum of uncertainties and disturbances in (7) is estimated with an adaptive law.

Case 1 Consider the first relation system in (13) with the control law (16) next to the adaptive control law (17). If the constants Λ, k, ρ and r are chosen so that $r > 0$, $k > 0$, $\rho > 1$ and $\Lambda > 0$, the error closed loop system is asymptotically stable and the value of x is exponential to the desired value x_des will like.

It should be noted that considering that the rank of $G(q)$ matrix in (5) is 2 for all equilibrium points. Therefore, the system is under excitation and therefore with two control inputs $[v_1, v_2]^T$, two state variables $[x, y]^T$ can follow the desired path $[x_{des}, y_{des}]^T$ controlled The sliding mode control law will be performed in a back-Stepping and in the following three stages:

First step: designing u_1 and θ in (8) and (9) so that $[x, y]^T$ follows $[x_{des}, y_{des}]^T$.

Second step: design δ in (10) so that θ converges to the desired value of θ_d which is calculated in the previous step.

Third step: design u_2 in (11) so that δ converges to the desired value δ_{des} which is calculated in the previous step.

***First step:* designing u_1 *and* θ**

First, consider changing the following variables:

$$v_1 = \cos(\theta)u_1, \qquad v_2 = \sin(\theta)u_1 \tag{12}$$

Considering (12), Eqs. (8)–(9) can be written as:

$$\dot{x} = v_1 + d_1, \qquad \dot{y} = v_2 + d_2 \tag{13}$$

As a result, the problem changes in such a way that we have to find the virtual inputs v_1 and v_2 in such a way that the states of x and y in (13) tend to the desired values of x_{des} and y_{des}. First, the design of the controller for variable x is done, and the process of designing the controller for y will be completely similar.

Considering the error as $\tilde{x} = x - x_{des}$, we choose the integral switching plane $s(t)$ as follows:

$$s(t) = \tilde{x} + \int_0^t \Lambda \tilde{x}(\tau)d\tau \tag{14}$$

where $s(t) \in R$ and $\Lambda > 0$ is an arbitrary positive number. In order to establish the sliding mode condition, the access rule is selected as follows

$$\dot{s}(t) = -rs - \rho\, sgn(s) \tag{15}$$

In this equation, the coefficients r and ρ are positive constants. Now, assuming that $\overline{\varepsilon}_1$ is the estimate of the upper band of d_1 in (8), i.e., ε_1, we propose the adaptive sliding mode control law as follows:

$$v_1 = \dot{x}_{des} - \Lambda\tilde{x} - rs - \rho\overline{\varepsilon}_1 sgn(s) \tag{16}$$

where $\overline{\varepsilon}_1$ is also obtained from the law of adaptive updating of the parameter:

$$\dot{\overline{\varepsilon}}_1 = k\rho s\, sgn(s) \tag{17}$$

In this relation, $k > 0$ is the gain of the relative term. By replacing relation (16) in the first relation (13), the derivative of relation (14) will be:

$$\dot{s}(t) = \dot{\tilde{x}} + \Lambda\tilde{x} = \dot{x} - \dot{x}_{des} + \Lambda\tilde{x} = -rs - \rho\, sgn(s) + d_1 \tag{18}$$

Stabilization of the control input introduced in (16) along with the adaptive law (17) is stated and proved in the form of the following theorem.

Theorem 1 *Consider the first system in relation (13) with the control law (16) next to the adaptive control law (17). If the constants Λ, k, ρ and r are chosen so that r > 0, k > 0, ρ > 1 and Λ > 0, the closed loop system error is asymptotically stable and the value of x will converge exponentially to the desired value x_{des}.*

Proof We choose the candidate Lyapunov function as below:

$$V = \frac{1}{2}s^2 + \frac{1}{2k}\tilde{\varepsilon}_1(t)^2 \tag{19}$$

where we have $\tilde{\varepsilon}_1(t) = \overline{\varepsilon}_1(t) - \varepsilon_1$. By deriving V with respect to time and substituting from relations (17) and (18) and with $\dot{\tilde{\varepsilon}}_1 = \dot{\overline{\varepsilon}}_1$, we have:

$$\begin{aligned}\dot{V} &= s\dot{s} + \frac{1}{k}\tilde{\varepsilon}_1\dot{\tilde{\varepsilon}}_1 = s(-rs - \rho\overline{\varepsilon}_1\, sgn(s) + d_1) + \frac{1}{k}\tilde{\varepsilon}_1\dot{\tilde{\varepsilon}}_1\\ &= -r|s|^2 - \rho\overline{\varepsilon}_1 s sgn(s) + sd_1 + \frac{1}{k}\tilde{\varepsilon}_1 k\rho s\, sgn(s)\\ &= -r|s|^2 - \rho\overline{\varepsilon}_1 s sgn(s) + sd_1 + (\overline{\varepsilon}_1 - \varepsilon_1)\rho s\, sgn(s)\end{aligned} \tag{20}$$

According to Assumption 1 and the relation $|d_1(t)| \leq \varepsilon_1$, its upper band can be written as follows:

$$\dot{V} \leq -r|s|^2 - \varepsilon_1\rho s\, sgn(s) + |s|\varepsilon_1 = -r|s|^2 + (1-\rho)\varepsilon_1|s| < 0 \tag{21}$$

Equation (21) shows that the condition for the derivative of the Lyapanov function to be negative is that $\rho > 1$. Therefore, because the Lyapunov function is positive definite and its derivative is negative definite, using the Lyapunov theorem (If we want to be more precise, we need to use Barbalat Lemma. In this case, according to Eq. (21) with the condition $\rho > 1$, the derivative of the Lyapunov function is negative semidefinite. By defining $\tilde{w}(t) = r|s|^2 + \mu|s|$ where $\mu = (1-\rho)\varepsilon_1$, and integrating both sides of the Eq. (21) we have $\infty > V(0) \geq \lim_{t\to\infty} \int_0^t \tilde{w}(\lambda)d\lambda$, and as a result, from Barballat's lemma, we can conclude that $\tilde{w}(t)$ and as a result $s(t)$ for t $\to \infty$ converge to zero.), the closed loop error system with controllers (16) and (17) is asymptotically stable and the error value tends to zero and as a result the value of x will tend exponentially to the desired value of x_{des}.

The design of control input v_2 for the second equation in (13) is also similar to (16). Therefore, v_2 and $\bar{\varepsilon}_2$ will be the estimates of the upper band of disturbances in relation with (9)

$$v_2 = \dot{y}_{des} - \Lambda_2 \tilde{y} - r_2 s_2 - \rho_2 \bar{\varepsilon}_2 sgn(s_2) \tag{22}$$

$$\dot{\bar{\varepsilon}}_2 = k_2 \rho_2 s_2 \, sgn(s_2) \tag{23}$$

Now, after calculating v_1 and v_2 in (16) and (22), using Eq. (12) the desired u_1 and θ values are calculated as follows:

$$\begin{aligned} u_1 &= \sqrt{v_1^2 + v_2^2} \\ \theta_{des} &= \tan^{-1}\left(\frac{v_2}{v_1}\right) \end{aligned} \tag{24}$$

Remark 1 Afterward, the value of u_2 should be designed in such a way that the value of θ in (10) tends to θ_{des} which has obtained in (24). For this purpose, first we find the value of δ in (10) in such a way that $\theta \to \theta_{des}$ and we call this value δ_{des} (desired δ). Finally, we find the value of u_2 in (11) in such a way that $\delta \to \delta_{des}$ is satisfied.

Second step: designing δ.

We consider relation (10) as follows:

$$\dot{\theta} = k(\delta - \varepsilon\theta)u_1 + d_3 = A\theta + B\delta + d_3 \tag{25}$$

where $A \triangleq -k\varepsilon u_1$, $B \triangleq k u_1$.

Where we have $\tilde{\theta} = \theta - \theta_{des}$ and Λ_3 is a positive constant. By choosing δ as below and applying it to the system (25), the value of θ will converge to θ_{des}.

$$\delta = \frac{1}{B}\left\{\dot{\theta}_{des} - \Lambda_3 \tilde{\theta} - A\theta - r_3 s_3 - \rho_3 \bar{\varepsilon}_3 \, sgn(s_3)\right\} \tag{26}$$

In this relation, the values of r_3, Λ_3 and ρ_3 are constant positive values and $\bar{\varepsilon}_3$ is the estimate of the upper uncertainty band of d_3 in (10) and is estimated with the following relative relation:

$$\dot{\bar{\varepsilon}}_3 = k_3 \rho_3 s_3 \, sgn(s_3) \tag{27}$$

In this relation, $k_3 > 0$ is the matching gain. And the sliding plane s_3 is also obtained from the following equation:

$$s_3(t) = \tilde{\theta} + \int_0^t \Lambda_3 \tilde{\theta}(\tau) d\tau \tag{28}$$

Substituting Eq. (26) into Eq. (25) leads to the following equation for the closed loop system:

$$\dot{s}_3(t) = -r_3 s_3 - \rho_3 \overline{\varepsilon}_3 sgn(s_3) + d_3 \tag{29}$$

Theorem 2 *Consider system (25). By applying the relation (26) in which the value of $\overline{\varepsilon}_3$ is obtained from the matching rule (27), if the control values are $r_3 > 0$, $\rho_3 > 1$, $\Lambda_3 > 0$ and $k_3 > 0$, the closed loop system is asymptotically stable and the value of θ will exponentially converge to the desired value of θ_{des}.*

Proof Same as Theorem 1.

***Third step*: designing u_2**

The last step of designing the adaptive-resistant sliding mode control law is to determine the value of u_2 in (11) in such a way that without any knowledge of the uncertainty value of d_4, the value of δ tends to the value obtained in (26).

The control input u_2 in relation (11) is proposed as below.

$$u_2 = \dot{\delta}_{des} - \Lambda_4 \tilde{\delta} - r_4 s_4 - \rho_4 \overline{\varepsilon}_4 \, sgn(s_4) \tag{30}$$

where $\overline{\varepsilon}_4$ is the estimate of the upper band of uncertainty d_4 and it is estimated from the following relative relation:

$$\dot{\overline{\varepsilon}}_4 = k_4 \rho_4 s_4 \, sgn(s_4) \tag{31}$$

In this relation, $k_4 > 0$ is the adaptive gain.

Theorem 3 *Consider system (11). By applying the relation (30) in which the value of $\overline{\varepsilon}_4$ is obtained from the matching rule (31), if the control values are $r_4 > 0$, $\rho_4 > 1$, $\Lambda_4 > 0$ and $k_4 > 0$, the closed loop system is asymptotically stable and the value of δ will tend exponentially to the desired value of δ_{des}.*

Proof Similar to Case 1.

With the design of u_1 and u_2, the process of designing the controller ends. By applying this controller, the tip of the probe takes the desired path $[x_{des}, y_{des}]^T$ in the presence of uncertainty and limited disturbances and without any information about these disturbances and uncertainties.

3.2 Upper Band of Unknown Functional Disturbances

The uncertainties in relation (6) with a coefficient are dependent on the system states, and the external disturbances are also dependent on the variables of the flexible probe state, such as the location and orientation of its tip. In fact, the forces applied by other body tissues as well as the elastic force of the skin at the time of injection are dependent on the location and orientation of the probe tip at the time of movement.

Therefore, considering its upper band as a number is not a complete assumption. In this section, we assume that the upper band of uncertainties and disturbances is not a fixed number but a linear function of the system states; so, we consider Assumption 1 as follows:

Assumption 2 We assume that the amount of external disturbances and uncertainties is limited as follows:

$$|d_i(t)| \leq \alpha_i \|q\| + \beta_i \quad i = 1, \ldots, 4 \tag{32}$$

where the values of α and β are unknown.

***First step:* designing u_1 and θ**

We consider relation (13) along with Assumption 2. Ignoring the index $i = 1, \ldots, 4$ for relations (8)–(11), suppose that $\widehat{\alpha}$ is the estimate of the value of α and $\widehat{\beta}$ is the estimate of the value of β. In this case, we consider the adaptive sliding mode control law in (16), as follows:

$$v_1 = \dot{x}_{des} - \Lambda\tilde{x} - rs - (\rho + \widehat{\beta} + \widehat{\alpha}\|q\|)sgn(s) \tag{33}$$

in which we consider the updating adaptive law of parameters as follows:

$$\dot{\widehat{\alpha}}(t) = k_1 \|q\| s \, sgn(s) \tag{34}$$

$$\dot{\widehat{\beta}}(t) = k_2 s \, sgn(s) \tag{35}$$

In this relation, $k_1, k_2 > 0$ are positive adaptive gains.

Theorem 4 *Consider the first system of relation in (13) along with Assumption 2 with the control law (33) along with the adaptive control law (34)–(35). If the constants r, k_1, k_2, ρ are chosen in such a way that $r > 0$, $k_1, k_2 > 0$, $\rho > 0$, the closed loop system error is asymptotically stable and the value of x will exponentially converge to the desired value of x_{des}.*

Proof By applying the new control law (33), relation (18) becomes the following:

$$\dot{s}(t) = -rs - (\rho + \widehat{\beta} + \widehat{\alpha}\|q\|)sgn(s) + d_1 \tag{36}$$

Now, consider the Lyapunov function candidate as follows:

$$V = \frac{1}{2}s^2 + \frac{1}{2k_1}\tilde{\alpha}(t)^2 + \frac{1}{2k_2}\tilde{\beta}(t)^2 \tag{37}$$

where $\tilde{\alpha}(t) = \hat{\alpha} - \alpha,\ \tilde{\beta}(t) = \hat{\beta} - \beta$. Taking derivative from V as follows:

$$\dot{V} = s\dot{s} + \frac{1}{k_1}\tilde{\alpha}\dot{\hat{\alpha}} + \frac{1}{k_2}\tilde{\beta}\dot{\hat{\beta}} \tag{38}$$

Considering that due to α and β being fixed, $\dot{\tilde{\alpha}} = \dot{\hat{\alpha}}$ and $\dot{\tilde{\beta}} = \dot{\hat{\beta}}$ by substituting the control law from (33) in (37) we have:

$$\dot{V} = s\Big(-rs - \Big(\rho + \hat{\beta} + \hat{\alpha}q\Big)sgn(s) + d_1\Big) + \frac{1}{k_1}\tilde{\alpha}\dot{\hat{\alpha}} + \frac{1}{k_2}\tilde{\beta}\dot{\hat{\beta}} \tag{39}$$

Applying the comparative rules from (34)–(35):

$$\begin{aligned}\dot{V} &= s\Big(-rs - (\rho + \widehat{\beta} + \widehat{\alpha}\|q\|)sgn(s) + d_1\Big) \\ &\quad + (\widehat{\alpha} - \alpha)\|q\|\,sgn(s) + \Big(\widehat{\beta} - \beta\Big)s\,sgn(s)\end{aligned} \tag{40}$$

which gives the result:

$$\dot{V} \le -r|s|^2 - \rho|s| + |s||d_1| - \alpha\|q\||s| - \beta|s| \tag{41}$$

Now, using Assumption 2 and the fact that $|d_1(t)| \le \alpha\|q\| + \beta$, we continue:

$$\dot{V} \le -r|s|^2 - \rho|s| + |s|(\alpha\|q\| + \beta) - \alpha\|q\||s| - \beta|s| = -r|s|^2 - \rho|s| \tag{42}$$

As a result, with the condition $\rho > 0$ and $r > 0$, because the Lyapanov function is positive definite and its time derivative is negative, it can be concluded that according to the Lyapanov stability theorem similar to before, the error values in the sequence on the desired path will be zero and this completes the proof. ■

***Second and third steps:* designing the desired δ and u_2**

Similar to what was mentioned in step 1, taking into account the Assumption 2 and considering that the upper band of disturbances is not a fixed value but a function of the soft modes of the probe with unknown coefficients, the optimal δ in (26) and u_2 in (30) will be as follows:

***Second step:* optimal δ**

$$\delta = \frac{1}{B}\Big\{\dot{\theta}_{des} - \Lambda_3\tilde{\theta} - A\theta - r_3 s_3 - \Big(\rho_3 + \hat{\beta}_3 + \hat{\alpha}_3\|q\|\Big)\overline{\varepsilon}_3\,sgn(s_3)\Big\} \tag{43}$$

$$\dot{\hat{\alpha}}_3(t) = k_3\|q\|s_3\, sgn(s_3) \tag{44}$$

$$\dot{\hat{\beta}}_3(t) = k_4 s_3\, sgn(s_3) \tag{45}$$

***Third step:* input u_2**

$$u_2 = \dot{\delta}_{des} - \Lambda_4\tilde{\delta} - r_4 s_4 - \left(\rho_4 + \hat{\beta}_4 + \hat{\alpha}_4\|q\|\right)\overline{\varepsilon}_4\, sgn(s_4) \tag{46}$$

$$\dot{\hat{\alpha}}_4(t) = k_5\|q\|s_4\, sgn(s_4) \tag{47}$$

$$\dot{\hat{\beta}}_4(t) = k_6 s_4\, sgn(s_4) \tag{48}$$

where $i = 3, \ldots, 6, k\_i > 0$ are positive gins in adaptive rules. The proof of the stability of these control rules is similar to the previous section, and it will not be repeated here.

4 Simulation Results

In this section, according to the robust adaptive sliding mode control design described in the previous section, the simulation results will be shown in the form of a numerical example. For this purpose, the optimal path of the probe is considered as follows. This path is the optimal path that is used in many reference articles to test the sequence on the optimal path of the probe [23]

$$\begin{aligned} x_{des} &= t \\ y_{des} &= \sin\left(\frac{t}{2}\right) \end{aligned} \tag{49}$$

For this purpose, the disturbance and uncertainty values are considered as a random function between 1 and 2 in addition to a sinusoidal function in each repetition and other controller parameters are also selected as follows:

$$\rho = 6, \Lambda = 5, r = 3, k_i = 2.5 \tag{50}$$

The probe coefficients are assumed as follows [17]:

$$\varepsilon = 7.2, k = 0.28 \tag{51}$$

The disturbance values are also considered depending on the probe modes as follows:

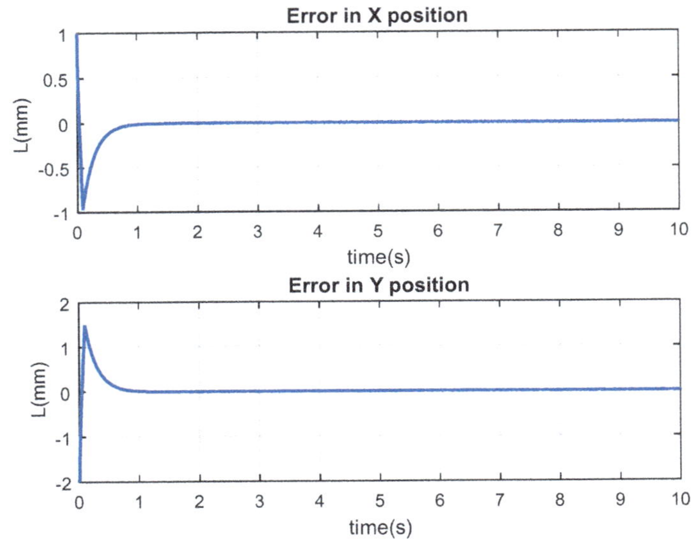

Fig. 2 Tracking error on the probe location in x and y directions

$$d = [2.5\|x\|, 2.5\|y\|]^T \tag{52}$$

Finally, the initial conditions $[x_0, y_0]^T = [1, -2]^T$ for the location of the probe tip and also $\hat{\beta}_1(0) = 3.5$ and $\hat{\beta}_2(0) = 4.3$ and also $\widehat{\alpha}_1(0) = 4$ and $\widehat{\alpha}_2(0) = 3$ are considered for adaptive laws.

By simulating in the software in MATLAB\Simulink, the results are shown in Figs. 2, 3, and 4.

As it can be seen from these figures, by applying the adaptive sliding mode control law in the presence of uncertainty and disturbance in (33), (43) and (46), the probe follows the desired path well and the error values in the direction x and y converge to zero, so we see in Fig. 2 that the error converges to zero in both directions.

The trajectory of the probe as well as the reference path in Fig. 3 confirm that the sinusoidal path is well followed when the movement starts. In Fig. 4, adaptive parameters in control laws (34) and (35) and adaptive laws related to control law v_2 are shown. As it can be seen from these figures, as expected with the convergence of the probe to the desired path, these parameters have converged to limited constant values.

Finally, in Fig. 5, it is clear that the sliding planes in control laws v_1 and v_2 have converged to zero in a limited time of about 1 s as expected. This issue is also true for the sliding surfaces related to the control laws of robust-adaptive sliding mode (43) and (46), which are omitted here because of the shortness.

In [23], a control rule based on model predictive control with a combination of feedback and feedforward parts is proposed for flexible bio inspired probe control, which is refrained from being detailed here. Considering the desired path

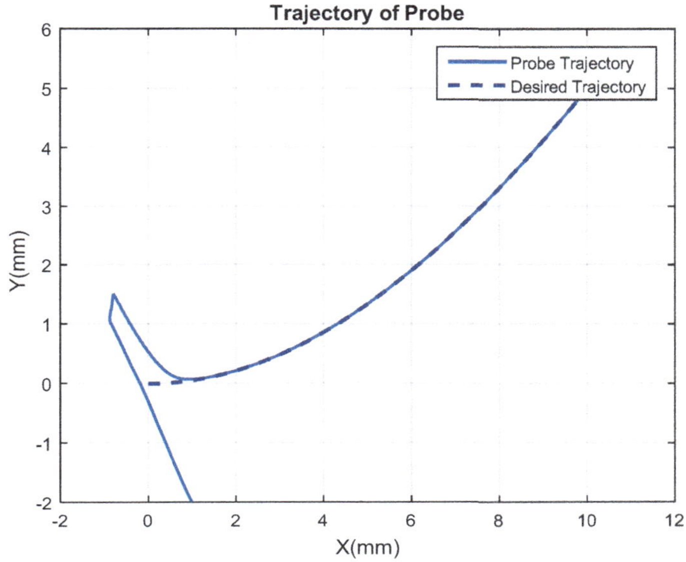

Fig. 3 Probe trajectory and desired reference path

as (49), the result of applying the adaptive-robust sliding mode control law in this paper compared to the predictive control method presented in [23], is shown in Figs. 6 and 7.

The set of results including a comparison with the reference paper in [23] in Figs. 6 and 7 clearly shows that in the control law of the robust-adaptive sliding mode by considering the upper band of disturbances as a function of the probe location, the effect of disturbances and uncertainties with a functional upper band is well compensated by the control law and the relevant adaptive laws.

5 Conclusion

In this paper, taking into account the model uncertainties in the kinematic model of the bio-flexible probe as well as the existence of external disturbances in the probe model, the problem of tracking the desired path in these flexible probes is considered. For this purpose, assuming the upper band of uncertainties and disturbances to be fixed and unknown, adaptive rules have been used to estimate this upper band and use this estimation in the robust-adaptive sliding mode control law. In the following, this upper band is considered as a function with unknown coefficients of the states norm of the system, and the corresponding sliding mode control law is developed by estimating these coefficients. The simulation results show the validation of the designed robust-adaptive sliding mode controllers.

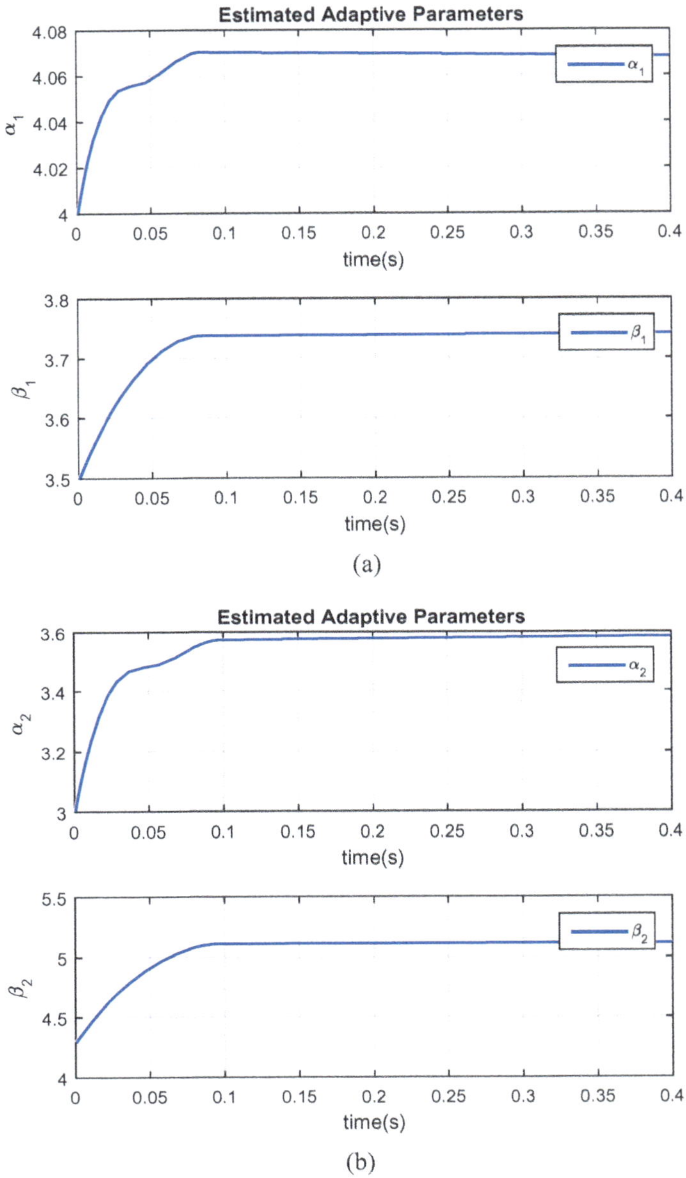

Fig. 4 Adaptive parameters in both control laws: **a** in the x direction and **b** in the y direction

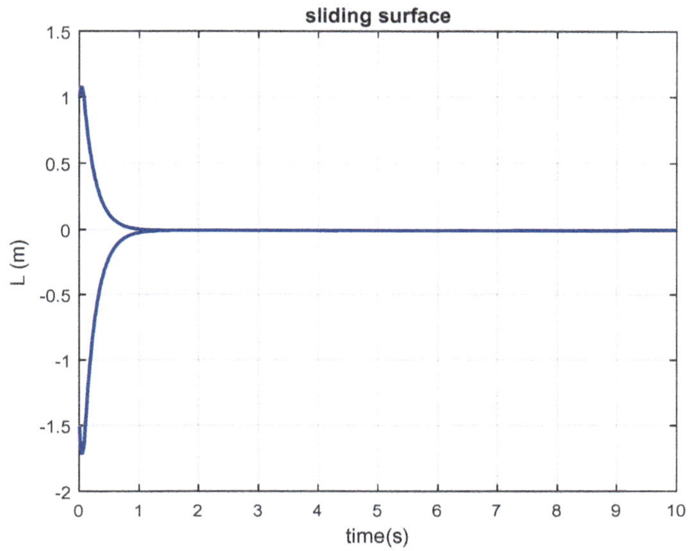

Fig. 5 Sliding surfaces without chattering phenomena

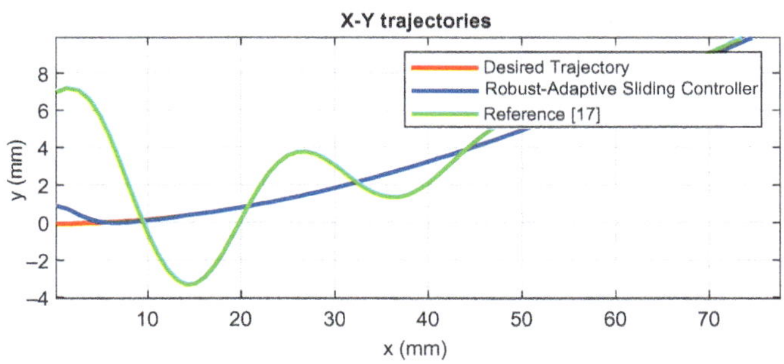

Fig.6 Probe path and desired reference path compared to the method presented in the article [23]

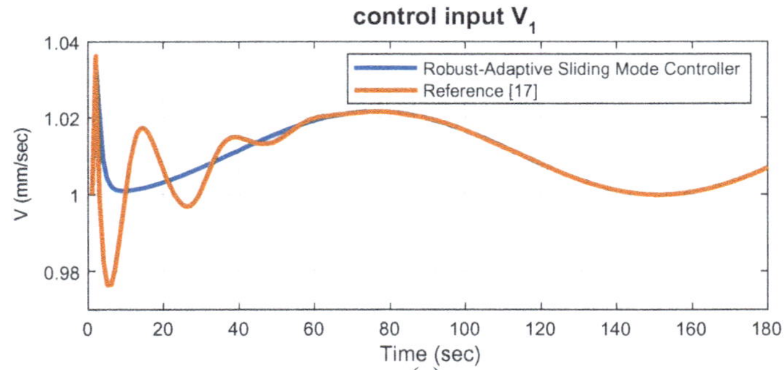

(a)

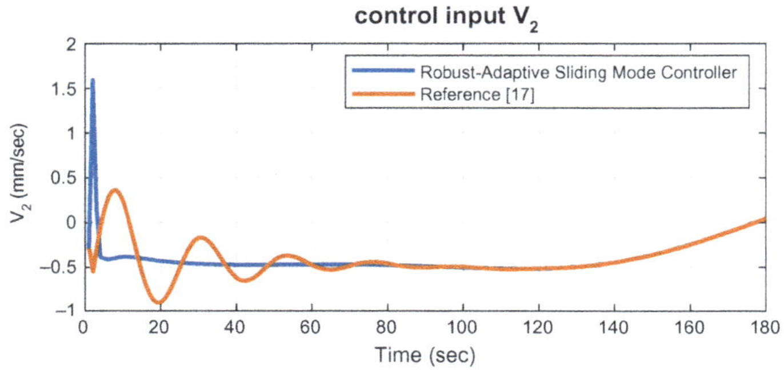

(b)

Fig. 7 Control inputs compared to the method presented in the article [23]. **a** v_1, **b** v_2

Author Contributions All authors contributed to the study conception and design. Material preparation, data collection and analysis were performed by all of the authors. The first draft of the manuscript was written by some and all authors commented on previous versions of the manuscript. All authors read and approved the final manuscript.

Conflict of Interest The authors declare that they have no conflict of interest.

Ethical Approval This study does not contain any studies with human participants or animals performed by any of the authors.

Informed Consent Informed consent was obtained from all individual participants included in the study.

References

1. Fuchs K (2010) Minimally invasive surgery. Endoscopy 34(2):154–159
2. Webster R, Kim J, Cowan N, Chirikjian G, Okamura A (2016) Nonholonomic modeling of needle steering. Int J Robot Res 25(5):509

3. Badie B, Brooks N, Souweidane M (2019) Endoscopic and minimally invasive microsurgical approaches for treating brain tumor patients. J Neurooncol 69(1):209–219
4. Alterovitz R, Goldberg K, Okamura A (2015) Planning for steerable bevel-tip needle insertion through 2d soft tissue with obstacles. In: Proceedings of ICRA, pp 1640–1645
5. Xu B, Zhou Ch, Ko SY (2018) Closed-loop planar fuzzy control system for a curvature-controllable steerable bevel-tip needle. Int J Control Autom Syst 16(5):2421–2431
6. Secoli R, Rodriguez F, Baena Y (2017) Adaptive path-following control for bio-inspired steerable needles. In: Conference: 6th IEEE international conference on biomedical robotics and biomechatronics (BioRob)
7. Zamiri S, Vahidian Kamyad A (2015) Robust trajectory tracking for a flexible probe in the presence of uncertainties and disturbance. J Math Comput Sci 14:108–123
8. Jushiddi MG et al (2020) Bevel angle study of flexible hollow needle insertion into biological mimetic soft-gel: simulation and experimental validation. J Mech Behav Biomed Mater 111:103896
9. Fallahi B et al (2019) An integrator-backstepping control approach for three-dimensional needle steering. IEEE/ASME Trans Mechatron 24(5):2204–2214
10. Scagilioni B et al (2019) Explicit model predictive control of a magnetic flexible endoscope. IEEE Robot Autom Lett 4(2):716–723
11. Cai Ch, et al (2020) Clinical flexible needle puncture path planning based on particle swarm optimization. Comput Methods Program Biomed 193:105511
12. Xu B, Ko SY (2020) 3D Feedback control using fuzzy logic for a curvature-controllable steerable bevel-tip needle. Mechatronics 68:202368
13. Hans S et al (2020) Robust control of a bevel-tip needle for medical interventional procedures. IEEE/CAA J Autom Sinica 7(1):244–256
14. Hans S, et al (2019) Continuous higher order sliding mode control of bevel-tip needle for percutaneous interventions. In: 28th IEEE international conference on robot and human interactive communication (RO-MAN), New Delhi, India
15. Hans S, et al (2020) Control of a flexible bevel-tipped needle using super-twisting controller based sliding mode observer, ISA Transactions (In Press)
16. Naderi Soorki M, Tavazoei MS (2018) Adaptive robust control of fractional-order swarm systems in the presence of model uncertainties and external disturbances. IET Control Theory Appl 12(7): 961–969 (2018)
17. Moghaddam TV, Nikravesh SKY, Khosravi MA (2019) Adaptive constrained sliding mode control of uncertain nonlinear fractional-order input affine systems. J Vibr Control 1–13. https://doi.org/10.1177/1077546319879484
18. Naderi Soorki M, Moghaddam TV, Emamifard A (2021) A new fast finite time fractional order adaptive sliding-mode control for a quadrotor. In: 7th international conference on control, instrumentation and automation (ICCIA)
19. de Vries M, Sikorski J, Misra S, van den Dobbelsteen JJ (2021) Axially rigid steerable needle with compliant active tip control. PLoS ONE 16(12):e0261089
20. Matheson E, Rodriguez Y, Baena F (2020) Biologically inspired surgical needle steering: technology and application of the programmable bevel-tip needle. Biomimetics (Basel) 5(4):68. PMID: 33339448
21. Konh B, Berkelman P, Karimi S (2020) Needle tip manipulation and control of a 3D steerable SMA-activated flexible needle. In: 2020 8th IEEE RAS/EMBS international conference for biomedical robotics and biomechatronics (BioRob), New York, pp 903–909
22. Hans S, Maria Joseph FO (2021) Control of a flexible bevel-tipped needle using super-twisting controller based sliding mode observer. ISA Trans 109:186–198
23. Ko SY (2010) Two-dimensional needle steering with a programmable bevel inspired by nature: modeling preliminaries. In: The 2010 IEEE/RSJ international conference on intelligent robots and systems, Oct 18–22, Taipei, Taiwan
24. Young Ko S, Rodriguez y Baena F (2012) Trajectory following for a flexible probe with state/input constraints: an approach based on model predictive control. Robot Auton Syst 60:509–521

Applied Mechatronics: A Case Study on Mathematical Modelling and Experimental Analysis of the Second Life Batteries

Adnan Mohamed Kaize, Farhad Salek, Aydin Azizi, Gordana Collier, and Shahaboddin Resalati

Abstract Second life batteries hold 80% of their capacity after being discarded from their use in EVs as they no longer meet the required specifications for the vehicle's usage. These batteries will provide terawatt-hours of capacity in the next few decades and are estimated to be at least half the price of a new Li-ion batteries. One of the main applications of the second life batteries are the buildings equipped with off-grid or grid-connected photovoltaic panels. In order to develop this technology for being widely used in energy storage systems, the techno-economic assessment is required. For techno-economic analysis of the second life battery energy storage systems, a validated mathematical model should be developed and used to estimate battery parameters. In this book chapter, the second life battery modelling using equivalent circuit model (ECM) is introduced. In addition, the battery testing protocol which is developed by high voltage energy storage (HVES) lab members is also presented. A validated mathematical model is described and integrated with a simple economic model for economic analysis of the energy storage system.

A. M. Kaize · A. Azizi (✉) · G. Collier
School of Engineering, Computing and Mathematics, Oxford Brookes University, Wheatley Campus, Oxford OX33 1HX, UK
e-mail: Aydin.azizi@brookes.ac.uk

A. M. Kaize
e-mail: 19139609@brookes.ac.uk

G. Collier
e-mail: gordanacollier@brookes.ac.uk

A. M. Kaize
AVL Powertrain UK Limited, Coventry, UK

F. Salek · S. Resalati
Faculty of Technology, Design and Environment, Oxford Brookes University, Oxford, UK
e-mail: F.salek@brookes.ac.uk

S. Resalati
e-mail: sresalati@brookes.ac.uk

© The Author(s), under exclusive license to Springer Nature Singapore Pte Ltd. 2023
A. Azizi (ed.), *Control Engineering in Mechatronics*, Emerging Trends in Mechatronics,
https://doi.org/10.1007/978-981-16-7775-5_3

1 Introduction

The transportation industry is one of the largest in the world and responsible for nearly 27% of total greenhouse gas emissions in the UK [1, 2], which has brought EVs being developed and encouraged in sale through subsidies and grants by governments due to their characteristic of being emission free. It is estimated that by 2030, the cumulative of EVs will reach 85 million [3]. These EVs have batteries that are imperative to the powertrain of the vehicle, and once it reaches a State of Health (SoH) of around 80%, it is no longer appropriate to drive EVs [4]. This creates a new market of Li-ion batteries that have 80% of their capacity available to be repurposed and used with a second life in other industries for various applications. Li-ion batteries are available in many different chemistries, each with their own benefits, however, majority of EV manufacturers opt for the NMC chemistry due to their high energy densities and low costs [3].

1.1 Application in Residential Buildings

Stationary Energy Storage is one of the most used and researched applications for SLB as it is less demanding and does not require high standards from the cells obtained from the EV battery [5]. Literature on this application don't entirely focus on the off-grid use of batteries in residential applications; however, they do provide vital information on the degradation of such batteries in these scenarios from End-of-life determination of these batteries to the effect of depth of discharge on the lifetime.

Many researchers agreed that the end of life of the SLB was determined to be at 60% SoH which is a 20% degradation from their starting SoH at 80% [6–8]. Based on the application, these batteries were able to last for varying amounts of time relying on several factors such as Depth-of-Discharge (DoD), Battery capacity and use-case. Cusenza et al. [6], showed that with a battery capacity of 9 kWh, the BESS would have a lifetime of 3 years, at 19 kWh, it lasted 6 years and by increasing it further, the battery lasted longer due to the higher capacities causing less stress on the battery, in an integrated load match analysis which did utilise the grid to power the residential building at times therefore, the BESS was not always under constant stress. It was also briefly mentioned in [9] that the second life cells tested in their analysis would last anywhere between 5.5 and 10.5 years in a maximum self-consumption PV residential scenario by estimating that a BESS undergoes 1 full equivalent cycle (FEC) per day and comparing that against their cell degradation profile.

1.2 Experimental Analysis

For development of an accurate model for the second life batteries, experimental data is required for validation and assessment of its accuracy. Two types of tests should be performed on the second life batteries in order to build a mathematical model; HPPC and degradation tests. The degradation of the second life batteries includes two mechanisms; calendar and cycle ageing. Calendar ageing is the ageing process of a battery while it is idle and comprises of all ageing processes that lead to the degradation of a battery cell independent of charge–discharge cycling. This form of ageing occurs from the development of passivation layers at the electrode–electrolyte interfaces [10]. The second form of ageing is the Cyclic ageing, which occurs due to the constant charging and discharging cycles that a cell undergoes. In addition to being influenced by the temperature and SoC levels, cyclic ageing is also influenced by the cut-off voltages of the charge and discharge, current rate, and the depth of discharge [11].

Different forms of reference performance tests are available to be conducted on a cell being tested during a cyclic ageing process. These tests are used to measure specific characteristics of the cell and the most common tests are Electrochemical Impedance Spectroscopy to determine capacity fading and impedance rise [12], Direct Current Internal Resistance (DCIR) test to find cell internal resistance [13], Pseudo-OCV to obtain OCV curve with respect to SoC of a cell and Hybrid Pulse Power Characterisation (HPPC) to define the dynamic performance characteristics of a battery by measuring the voltage drop resulting from a square-wave current load [14].

1.3 Modelling

Modelling of the BESS is mainly done using empirical models, such as equivalent circuit models (ECMs), which are simplified models which predicts cell parameters using resistor–capacitor pairs are able to predict overall battery behaviour without the need for complex physical mechanisms being modelled, they require much less computational effort [15, 16]. These models require experimental data to be used to validate the model and the quality of the model is directly dependent on the parameterisation data of the cell over various SoC levels and temperatures obtained experimentally.

Figure 1 shows a schematic design of an ECM with 2RC branches. The controllable voltage of ECM is defined by the 'Voc', which is the open circuit voltage of the Li-ion cell which would vary non-linearly with SoC. 'R0' is the internal resistance of the cell describing the connection resistance of the cell. 'R1' and 'R2' are the polarisation resistances of the cell alongside 'C1' and 'C2' which denote the polarisation capacitance. 'I' represents the current flowing through the cell, whereas

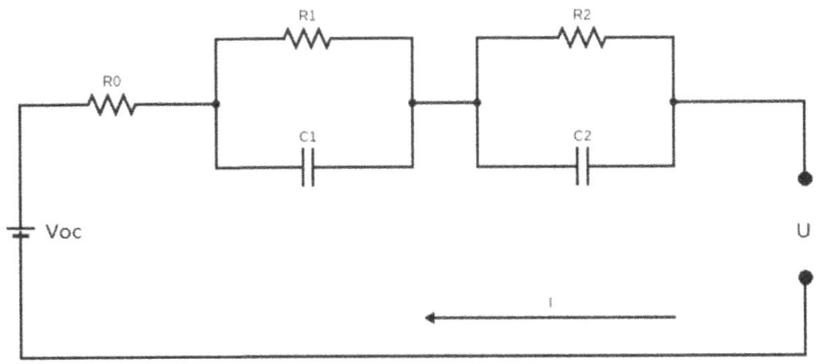

Fig. 1 Equivalent circuit model of Li-ion cell

'U' represents the terminal voltage of the cell, both of which are measurable using appropriate sensors [17].

2 Theoretical Analysis of the Second Life Batteries

Figure 2 shows a flow chart created, detailing the tasks required from carrying out experimental analysis to conducting the numerical analysis. These tasks were drafted to ensure that all objectives could be met in a timely manner and was decided based on the literature review carried out. The methodology can be divided into experimental and numerical stages.

2.1 Experimental Methodology

The experimental analysis is extremely important in the project battery modelling process as the data obtained from the experimental tests would be used to validate the numerical model, which is imperative to achieving the aim of the project. The HVES Lab at Oxford Brookes University was utilised to carry out the experiments. The methodology is detailed in the following sub-sections.

2.1.1 Equipment

A Binder KB115 Thermal Chamber was used to keep the battery cell at a constant temperature and a cyclic electrical load was applied to age the cell using an ARBIN LBT-21084-HC cell cycler. A Gamry Interface 5000P EIS Potentiostat was used to conduct the EIS testing on the cell. Figure 3 provides a picture of the apparatus.

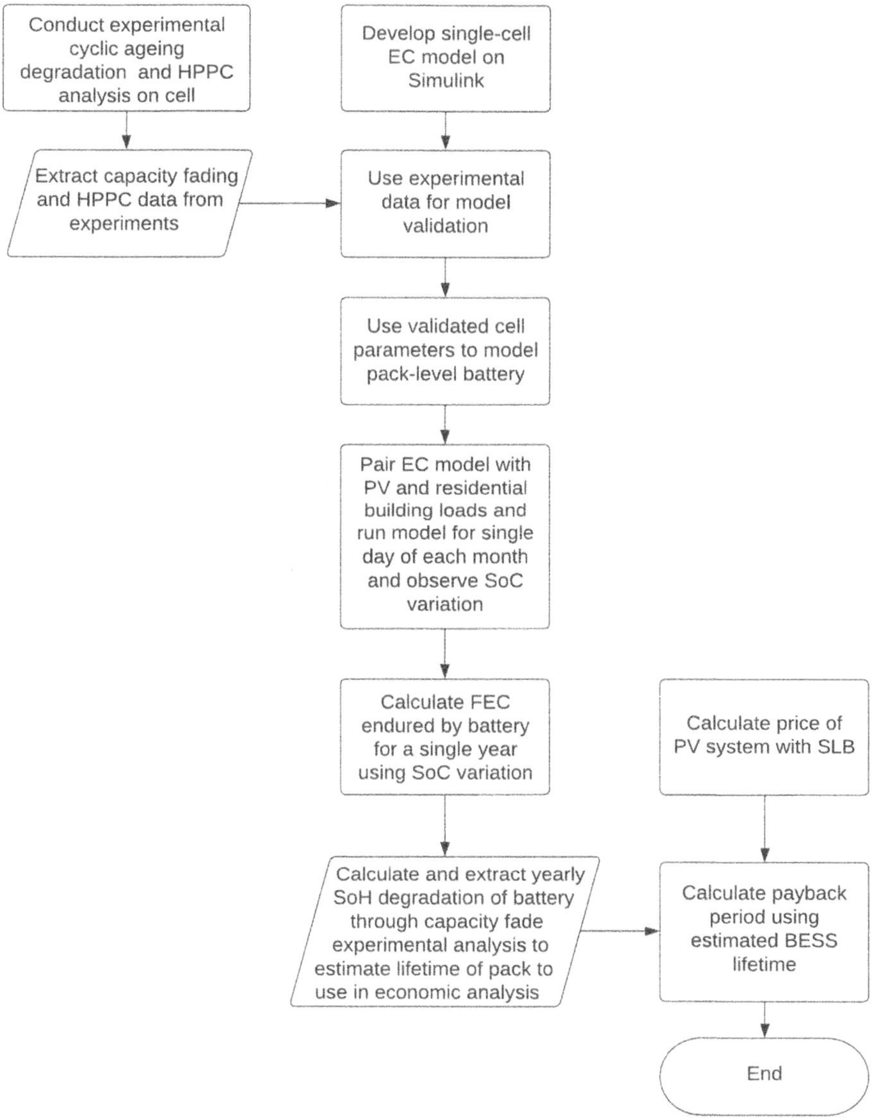

Fig. 2 Flow chart

All testing was completed in an environmental chamber using adequate fixtures to ensure safety of the lab and integrity of the experiment.

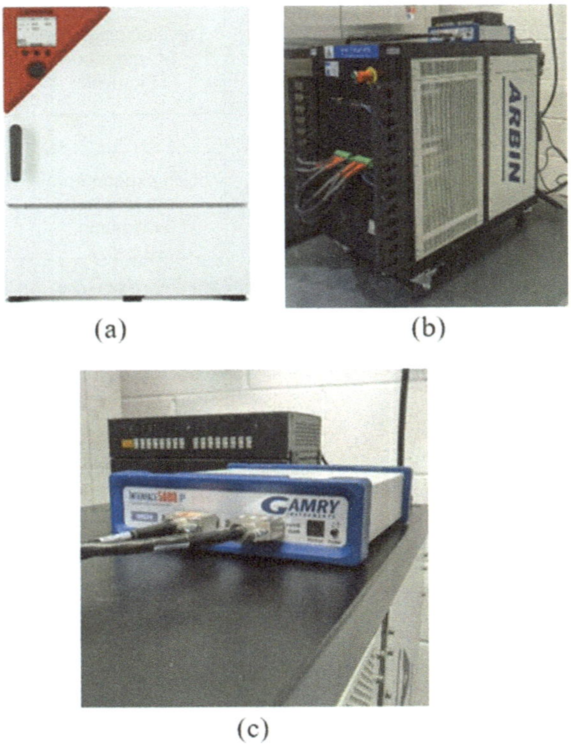

Fig. 3 **a** Thermal chamber, **b** Cell cycler, and **c** EIS Potentiostat equipment at HVES lab

2.1.2 Testing Limitations and Conditions

A Sony VTC6 cell was used for the experimental procedure. The cell had previously undergone a cyclic degradation process and was at a SoH of 86%, with a cell capacity of 2.68 Ah. Further cell specifications are shown in Table 1.

Table 1 Rated cell specifications

Cell	Chemistry	Nominal Voltage (V)	Initial AC impedance (mΩ)	Initial DC resistance (mΩ)	Nominal capacity (Ah)	Energy density (Wh/kg)	Power density (W/kg)
Sony VTC6	NMC	3.7	8–18	N/A	3	232	4056
Standard charge	Max voltage (V)	Max charge current (A)	Max discharge current (A)				
CC/CV	4.2	5	30				

a)
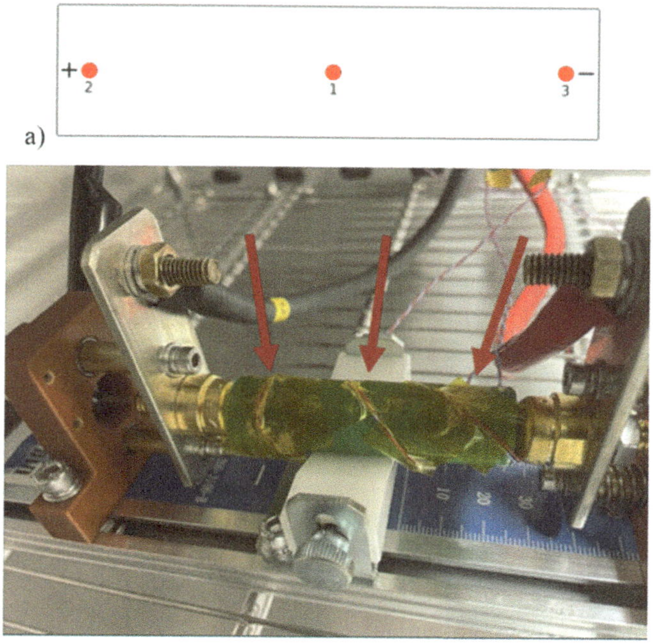

b)

Fig. 4 **a** Schematic thermocouple placement on cylindrical cells and **b** Experimental placement

Type-T thermocouples were used to sense the temperatures of the cell. These thermocouples were connected with the cell cycler to acquire the temperature data. Figure 4 shows the schematic arrangement of the thermocouples on the cylindrical cell used for this experiment as well as how it was connected in the experimental procedure. Thermal paste was used on the locations to ensure that the measurements were consistent and as accurate as possible, and the paste was coupled with Kapton tape to hold the thermocouples in place.

2.1.3 Overall Testing Protocol

The individual test definitions below were used to attain the full characterisation of the cell. As the degradation analysis of the cylindrical battery cell in a single environmental temperature is focused, the cycle ageing tests were designed to evaluate the cell degradation behaviour. All the tests were performed at 30 °C. Internal resistance measurements used a current amplitude of 250 mA. The frequency used for these internal resistance measurements was 10 Hz.

To perform Pseudo-OCV test, a discharge and charge of C/25 was completed with a minimum 1-h rest in between to acquire a Pseudo-OCV curve at the constant temperature of 30 °C.

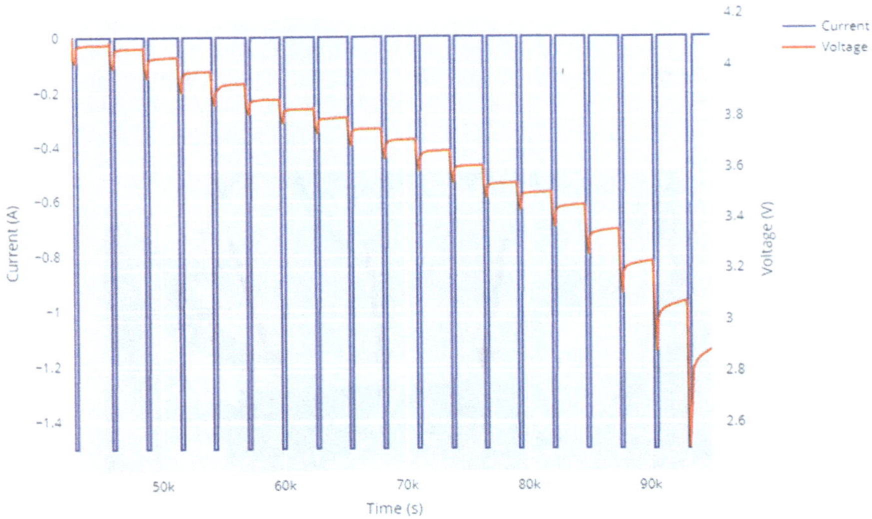

Fig. 5 HPPC pulse data with voltage response

A galvanostatic (constant current) EIS test was performed from 1 MHz to 1 kHz with 5% SoC increments and 8 points per decade. The test was completed at 30 °C with a current amplitude of 250 mA.

The HPPC tests were done at a 100% SoH while it was being degraded in its first life. The detail of the test is given below.

The hybrid pulse power characterisation [18] test was completed to determine the pulse power capabilities of the cell at 25 °C. These tests were completed with the following conditions and with pulses of 5% SOC increments. The pulse profile used is shown in Fig. 5 along with the voltage response of the cell. A 1.5 A discharge current pulse was used.

A constant cell cycling of an endurance drive cycle that included 22 laps with a 3-min rest halfway through, followed by a 30-min rest and then a standard charge was used with an ambient temperature of 30 °C maintained, Drive cycle specification was created by the Oxford Brookes University HVES lab members [19] and shown in Fig. 6.

The cell was cycled for 200 additional cycles, after already being cycled 200 times during its first life. Every ten cycles a reference performance test was completed including an EIS test, internal resistance test, and a static capacity at standard charge and C/2 discharge. A pseudo-OCV curve is acquired every 40 cycles in addition to the 10-cycle RPT.

The cylindrical cell alongside the cell holder was placed inside the Thermal chamber which was used to maintain a constant temperature of 30 °C as shown in Fig. 7. The inner black and red connections were used to connect the cell with the cycler and the outer crocodile clips were used to take Voltage measurements.

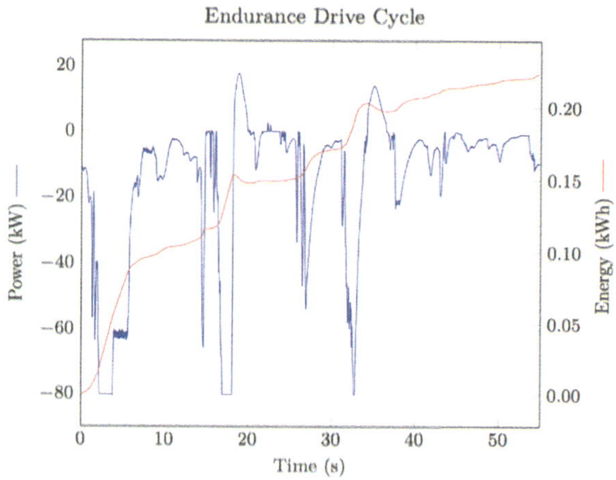

Fig. 6 Endurance drive cycle—power and used energy versus time [19]

Fig. 7 Cylindrical cell inside thermal chamber with connections to cell cycler

Finally, once the cell had been placed inside the chamber and all connections been made, the chamber was closed with a temperature of 30 °C set. A 2-h wait was required for the temperature in the chamber to reach the desired temperature and then the cyclic degradation began.

2.2 *Mathematical Modelling*

Due to the nature of a SLB undergoing constant charging and discharging cycles through photo-voltaic charging and demand from the residential building respectively, as well as the nature of the project with respect to time, only cyclic ageing was modelled numerically in this study.

The numerical methodology consisted of designing an Equivalent Circuit Model (ECM) of a battery cell with a controlled current source using Simscape Electrical on Simulink. This cell-level model was then validated using experimental data obtained from the experimental HPPC analysis done on the Sony VTC6 cell.

Once the validation was completed, the model was further extended to a pack level and standard UK residential building loads along with the power output from a PV system were applied to the model to monitor the SoC variation for representative days of each month to estimate the SoH decline of the battery. The decline in the capacity of the battery pack was used as an indicator of the SoH. Further details of the specification of the PV system, battery pack capacity and loads are provided in the subsequent sub-sections.

To validate an individual cell with the experimental data, a cell-level model was designed using the Simscape Electrical toolbox found in MATLAB 2022a. Figure 9 shows the model designed for this.

As shown by Fig. 8, the model consists of the following components to perform the parameter estimation for validating the model:

- Battery (Table-Based)—Used to model the cell based on tabulated characteristics as functions of state of charge (SoC) and optionally temperature. It allows the user to parameterise based on the no-load voltage and terminal resistance with respect to State of Charge, cell capacity, cell dynamic modelling, as well as the fade characteristics of the cell.
- Controlled Current Source—Used to represent an ideal current source that is powerful enough to maintain the specified current through it, regardless of the voltage across it.
- Voltage Sensor—Used to measure the voltage across a part.

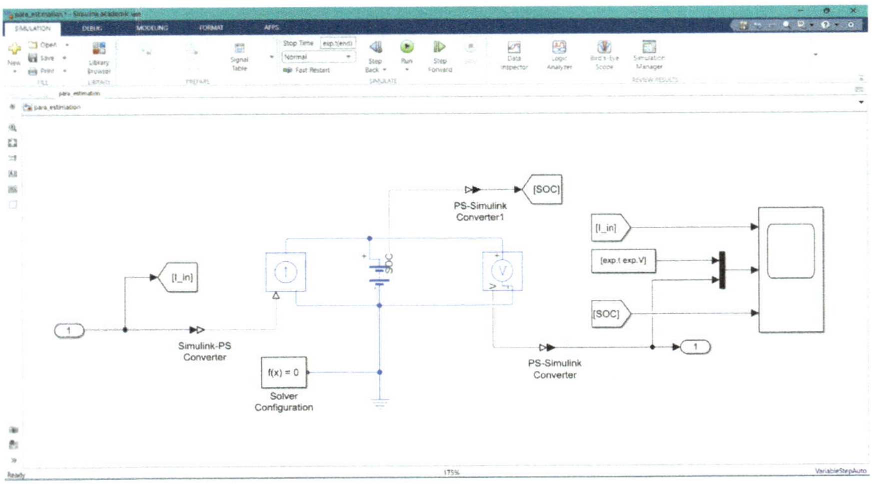

Fig. 8 Cell-level model for validation

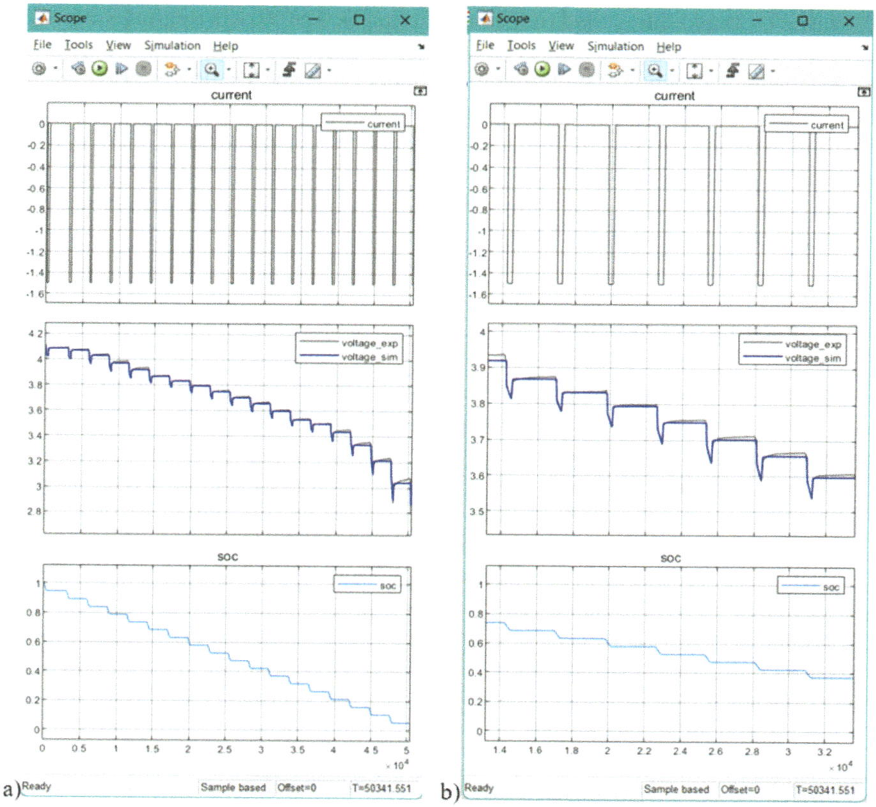

Fig. 9 **a** Parameter estimation results on HPPC data. **b** Zoomed in model

Using this model, the cell was parameterised and prepared for fitting using the parameter estimation toolbox found in Simulink.

2.2.1 Fitting ECM Parameters

The model was fitted using the experimental data obtained to have a validated model that behaves similarly to the real-life behaviour of the cell. This was done by using the HPPC test data of the cell which measured the voltage drop resulting from a square-wave current load. The isothermal pulse discharge HPPC test data at 25 °C used to validate the model is shown in Fig. 5. The Parameter Estimator toolbox within Simulink was used to fit the model parameters with the experimental data. As the dynamic performance characteristic was studied in HPPC tests, the different parameters selected to be estimated were, No Load Voltage (Em), Internal Resistance (R0), First polarisation resistance (R1), first time constant (tau1), second polarising resistance (R2) and the second time constant (tau2). The "Trust-Region-Reflective" algorithm alongside the "Nonlinear least squared" methodology was used to perform the parameter estimation. The fitting of the model was successfully carried out and the estimated parameters showed a very close, almost identical resemblance to the experimental data giving an R-squared value of 0.999. The results of the parameter estimation are shown in Fig. 9.

Once fitting was completed, the parameters obtained from the estimation were inputted into the cell model in Simulink.

2.2.2 Pack Level Modelling

Once all PV-system and battery parameters have been decided and the loads into and out of the system were accounted for, the battery model was ready to be constructed using the validated cell parameters that were generated from parameter estimation.

To design a battery pack with the specifications of 48 V and 13 kWh, the Sony VTC6 cell voltage and capacity at 80% SoH need to be considered. As the cell has a voltage of 4 V and a capacity of 2.46 Ah at 80% SoH, the battery would require a capacity of 270.8 Ah to get a rating of 13 kWh at 48 V. To achieve this, the cell was designed with 10 modules connected in parallel, each module consisting of 11 parallel strings with 12 cells in each string. Figure 10a and b show the module and pack configuration of the battery pack respectively. With this configuration, the battery pack achieves a voltage of 48 V and a capacity of 270.6 Ah, giving it a total energy capacity of 12,988.8 kWh.

2.2.3 Battery Load Simulation

The full battery model is shown below in Fig. 11. This model consisted of the battery pack connected to a power source and a dynamic load to charge and discharge the pack respectively. A scope was added into a single cell's SoC port to monitor the battery pack's SoC during the charging and discharging process. The pack was assumed to act homogenously in terms of SoC, where the SoC in all cells are equal throughout the process.

The following components were used in the model:

1. Dynamic Load—This block implements a dynamic power load on the battery. Input into this block was the resultant power supplied to the house by the battery with respect to time, after taking PV self-consumption into account.
2. Voltage Sensor—This block was used to monitor the voltage of the battery pack.
3. Electrical Power Source—This component is used to input power into a system. It works by dividing the power input from the PVGIS tool by the battery voltage state and using a controlled current source to input a varying current into the system based on the battery's voltage.

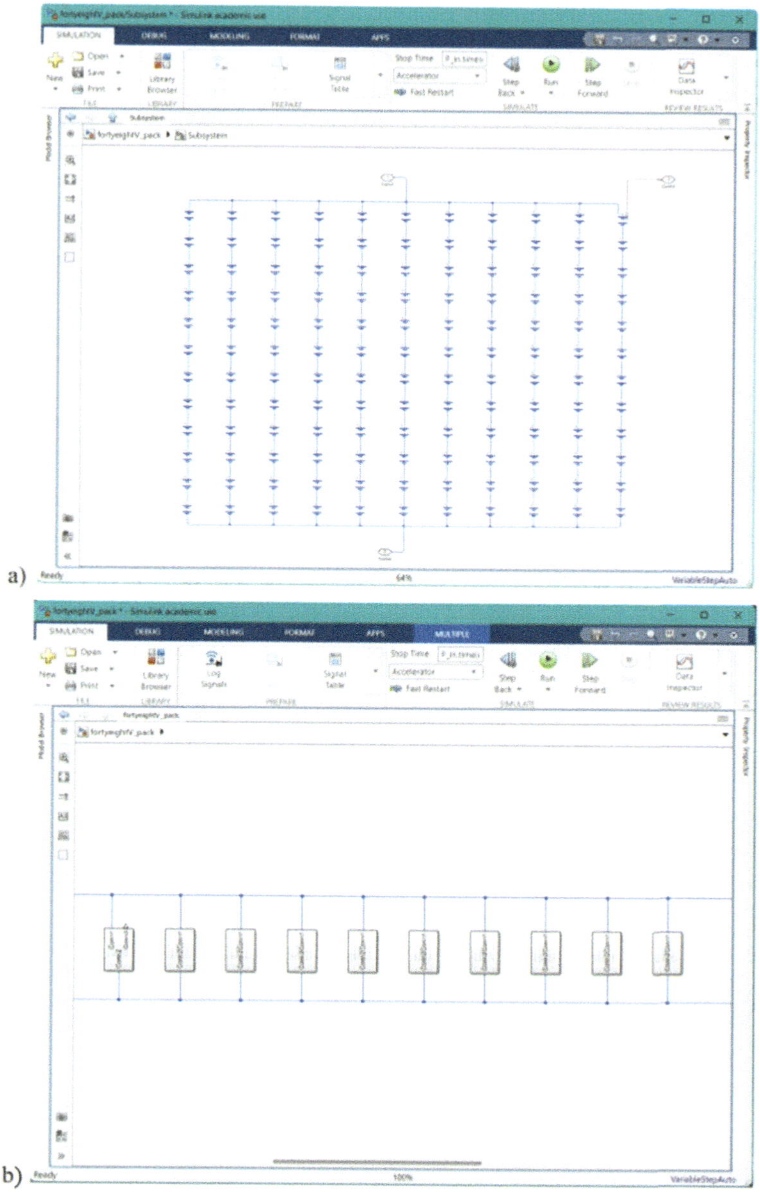

Fig. 10 **a** Battery module configuration with 11 parallel strings of 12 cells connected in series and **b** the module arrangement in pack

Fig. 11 Battery model used to estimate SoC for SoH estimation

The power in and out of the system were defined for each simulation, and a 40% SoC was assumed to be remaining in the battery from the previous day at the start of each day to manage loads before sunrise. Once each simulation was run, the SoC variation with time is obtained and the FECs per day were estimated using the equation below, taken from Motapol et al. [20], where the profile was converted into the number of cycles where the SoC of the battery was 100% at the start and end of the cycle.

$$N_{eq} = 0.5\left(1 - \frac{DoD_1}{DoD_2}\right) + 0.5\left(1 - \frac{DoD_3}{DoD_2}\right) \quad (1)$$

For example, if a battery was discharged from 80% SoC (20% DoD) to 60% SoC (40% DoD), and then charged back up to 70% SoC (30% DoD), the number of FECs would be:

$$N_{eq} = 0.5\left(1 - \frac{20\%}{40\%}\right) + 0.5\left(1 - \frac{30\%}{40\%}\right) \quad (2)$$

$$N_{eq} = 0.375$$

This method was used to estimate the FECs experienced by the battery pack for a single representative day of each month, and then the total number of cycles experienced by the battery pack in a year was calculated with the assumption that the number of cycles in the representative day of each month would be the same for the rest of the days in the month. With this data, the lifetime of the BESS in this scenario could be estimated by comparing the FECs to that of the experimental data obtained from the degradation tests, and the time required to reach EoL could be estimated.

2.2.4 State of Health Estimation

As the data obtained from the experimental tests were not run till the end of life of a second-life cell, which was seen to be at 60% SoH in the literature review, predictive methods were used to extend the cyclic degradation data and extend the range of capacity fade for further analysis. This was done through a Standard Neural Network designed using the "Keras deep learning API" in Jupyter Notebooks. Standard Neural Networks work by using the provided data to train a model which would be used to predict the behaviour of the model with additional inputs. They start by taking a single data point with random numbers for the weights and makes a prediction, based on how accurate the prediction is, the neural network adjusts the weights and moves on to the next point.

The Neural Network consisted of an input layer taking a single feature, which was the cycle index. The input layer was followed by 8 additional hidden layers, including batch normalisation and finally, an output layer which consisted of the prediction, in this case, the capacity of the cell. The data was divided into two sets, training, and testing with an 80–20 split respectively. The training set was used to train the model with the input data and the testing set was used to validate the model. A root mean squared analysis was used to test the accuracy metric.

References

1. Alimujiang A, Jiang P (2020) Synergy and co-benefits of reducing CO_2 and air pollutant emissions by promoting electric vehicles—a case of Shanghai. Energy Sustain Dev 55:181–189
2. Transport and environment statistics: Autumn 2021. https://www.gov.uk/government/statistics/transport-and-environment-statistics-autumn-2021/transport-and-environment-statistics-autumn-2021
3. Haram MHSM, Lee JW, Ramasamy G, Ngu EE, Thiagarajah SP, Lee YH (2021) Feasibility of utilising second life EV batteries: applications, lifespan, economics, environmental impact, assessment, and challenges. Alex Eng J 60(5):4517–4536
4. Tong S, Fung T, Klein MP, Weisbach DA, Park JW (2017) Demonstration of reusing electric vehicle battery for solar energy storage and demand side management. J Energy Storage 11:200–210
5. Zhu J, Mathews I, Ren D, Li W, Cogswell D, Xing B et al (2021) End-of-life or second-life options for retired electric vehicle batteries. Cell Rep Phys Sci 2(8):100537
6. Cusenza MA, Guarino F, Longo S, Mistretta M, Cellura M (2019) Reuse of electric vehicle batteries in buildings: an integrated load match analysis and life cycle assessment approach. Energy Build 186:339–354
7. Assunção A, Moura PS, de Almeida AT (2016) Technical and economic assessment of the secondary use of repurposed electric vehicle batteries in the residential sector to support solar energy. Appl Energy 181:120–131
8. Mathews I, Xu B, He W, Barreto V, Buonassisi T, Peters IM (2020) Technoeconomic model of second-life batteries for utility-scale solar considering calendar and cycle aging. Appl Energy 269:115127
9. Braco E, San Martín I, Berrueta A, Sanchis P, Ursúa A (2020) Experimental assessment of cycling ageing of lithium-ion second-life batteries from electric vehicles. J Energy Storage 32:101695
10. Keila P, Schustera S, Wilhelma J, Travia J, Hauserb A, Karla R, et al (2016) Calendar aging of lithium-ion batteries I. Impact of the graphite anode on capacity fade. J Electrochem Soc 163:A1872–A80
11. Barcellona S, Piegari L (2020) Effect of current on cycle aging of lithium ion batteries. J Energy Storage 29:101310
12. Middlemiss LA, Rennie AJ, Sayers R, West AR (2020) Characterisation of batteries by electrochemical impedance spectroscopy. Energy Rep 6:232–241
13. Noelle DJ, Wang M, Le AV, Shi Y, Qiao Y (2018) Internal resistance and polarization dynamics of lithium-ion batteries upon internal shorting. Appl Energy 212:796–808
14. Barai A, Uddin K, Dubarry M, Somerville L, McGordon A, Jennings P et al (2019) A comparison of methodologies for the non-invasive characterisation of commercial Li-ion cells. Prog Energy Combust Sci 72:1–31
15. You HW, Bae JI, Cho SJ, Lee JM, Kim S-H (2018) Analysis of equivalent circuit models in lithium-ion batteries. AIP Adv 8(12):125101
16. Tian N, Wang Y, Chen J, Fang H (Eds) (2017) On parameter identification of an equivalent circuit model for lithium-ion batteries. In: 2017 IEEE conference on control technology and applications (CCTA). IEEE
17. Zhang L, Peng H, Ning Z, Mu Z, Sun C (2017) Comparative research on RC equivalent circuit models for lithium-ion batteries of electric vehicles. Appl Sci 7(10):1002

18. Salek F, Azizi A, Resalati S, Henshall P, Morrey D (2022) Mathematical modelling and simulation of second life battery pack with heterogeneous state of health. Mathematics 10(20):3843
19. Planden B, Lukow K, Colliera G, Morreya D (2021) Battery testing consortium: a framework for improvements in high-power battery design. http://github.com/Oxford-Brookes-HVES/BTC
20. Motapon SN, Lachance E, Dessaint L-A, Al-Haddad K (2020) A generic cycle life model for lithium-ion batteries based on fatigue theory and equivalent cycle counting. IEEE Open J Ind Electron Soc 1:207–217
21. Ambrose H, Kendall A, Slattery M, Steckel T. Battery second-life: unpacking opportunities and barriers for the reuse of electric vehicle batteries. https://pdfslide.net/documents/battery-second-life-unpacking-opportunities-and-2020-05-27-battery-second-life.html?page=5
22. BNEF (2021). https://about.bnef.com/blog/battery-pack-prices-fall-to-an-average-of-132-kwh-but-rising-commodity-prices-start-to-bite/
23. Kelleher Environmental (2019) Research study on reuse and recycling of batteries employed in electric vehicles: The technical, environmental, economic, energy and cost implications of reusing and recycling EV batteries. Energy API
24. Melin HE (Ed) (2018) The lithium-ion battery end-of-life market—a baseline study. World Economic Forum, Cologny, Switzerland
25. Jaiswal A (2017) Lithium-ion battery based renewable energy solution for off-grid electricity: a techno-economic analysis. Renew Sustain Energy Rev 72:922–934
26. Al-Wreikat Y, Attfield EK, Sodré JR (2022) Model for payback time of using retired electric vehicle batteries in residential energy storage systems. Energy 259:124975
27. Beggs C (2010) Energy: management, supply and conservation. Routledge

Embracing Digital Technologies in the Pharmaceutical Industry

Reza Ebrahimi Hariry and Reza Vatankhah Barenji

Abstract The pharmaceutical sector is vital to the healthcare system. Without it, drug discovery, development, and distribution would be impossible. When we say "the pharmaceutical industry, "we refer to research, development, and distribution. Pharmaceutical and healthcare companies are under great pressure to innovate via research and development and to stay compliant at all costs considering recent technological advancements. Unfortunately, this pressure negatively impacts the effectiveness of clinical trials, operational efficiency, sales, and ultimately the development of this sector. Utilizing digital technology may provide substantial benefits for the global pharmaceutical industry. In this book chapter, we are reviewing the digital technologies that affect the productivity of the pharmaceutical industry.

1 Introduction

The pharmaceutical industry is undergoing a major transformation, driven by advances in digital technology. The rise of digital technologies has brought about a multitude of benefits for the pharmaceutical industry, including improved efficiency, accuracy, and cost-effectiveness. In recent years, there has been a significant increase in the adoption of digital technologies in the pharmaceutical industry, from drug discovery and development to distribution and patient care. The pharmaceutical industry has been relatively slow to adopt new technologies compared to other industries, such as finance or retail. However, this is changing, as pharmaceutical companies are recognizing the potential benefits that digital technologies can bring. Digital technologies are having a profound impact on the way that pharmaceutical company's approach to research and development, clinical trials, and patient care [1].

R. E. Hariry (✉)
Department of Pharmacology and Health Technology, Faculty of Pharmacy, Girne American University, Kyrenia, Cyprus
e-mail: rezahariry@gau.edu.tr

R. V. Barenji
Department of Engineering, School of Science and Technology, Nottingham Trent University, Nottingham NG11 8NS, UK

© The Author(s), under exclusive license to Springer Nature Singapore Pte Ltd. 2023
A. Azizi (ed.), *Control Engineering in Mechatronics*, Emerging Trends in Mechatronics,
https://doi.org/10.1007/978-981-16-7775-5_4

One of the key benefits of digital technologies in the pharmaceutical industry is improved efficiency. Digital technologies have the potential to streamline many of the processes involved in drug discovery and development, from the identification of new drug targets to the optimization of drug candidates to the analysis of clinical trial data. By automating many of these processes, pharmaceutical companies can save time and resources, while increasing the speed and accuracy of their research and development activities. Another benefit of digital technologies in the pharmaceutical industry is improved accuracy. With the use of digital tools, such as computer-aided drug design and simulation software, pharmaceutical companies can now make more informed decisions about which drug candidates to pursue, and how to optimize their properties for maximum efficacy and safety. In addition, digital technologies are enabling pharmaceutical companies to better track and analyze clinical trial data, which can help to improve the accuracy of results and speed up the regulatory approval process [2].

In addition to improved efficiency and accuracy, digital technologies are also helping to reduce costs in the pharmaceutical industry. By automating many of the processes involved in drug discovery and development, pharmaceutical companies can reduce the need for manual labor, which can lead to significant cost savings. Furthermore, digital technologies are enabling pharmaceutical companies to reduce the time and resources required for clinical trials, which can also result in significant cost savings. Despite the many benefits of digital technologies in the pharmaceutical industry, there are also challenges that must be overcome. One of the biggest challenges is ensuring the security and privacy of patient data. With the increasing use of digital technologies in the pharmaceutical industry, there is a growing risk of data breaches and cyber-attacks. This is a particular concern for clinical trial data, which can be highly sensitive and confidential [3].

Another challenge is ensuring the reliability and accuracy of digital technologies. As digital technologies become more widespread in the pharmaceutical industry, it is important to ensure that they are reliable and accurate and that they meet the highest standards of quality and safety. This requires ongoing investment in research and development, as well as robust quality control processes to ensure that digital technologies are fit for purpose. Despite these challenges, there is no doubt that digital technologies are having a profound impact on the pharmaceutical industry. From improved efficiency and accuracy to reduced costs and better patient care, the benefits of digital technologies are clear. As such, pharmaceutical companies must embrace digital technologies, and work to overcome the challenges that are associated with their adoption, to stay ahead of the curve and remain competitive in the fast-paced and rapidly evolving world of pharmaceuticals [4].

The aim of this paper is to review the potential benefits of digital technologies in the pharmaceutical industry. The review will focus on the ways in which digital technologies are improving efficiency, accuracy, and cost-effectiveness in the pharmaceutical industry, and the challenges that must be overcome to fully realize these benefits. The ultimate goal of this review is to provide a clear understanding of the potential benefits of digital technologies in the pharmaceutical industry and to provide insights into how these technologies can be harnessed to improve the development and delivery of drugs and enhance patient care.

2 Internet of Things

The Internet of things (IoT) refers to the network of physical items or 'Things' implanted with electronic software, sensors, and network connection, allowing them to gather and share data [5]. In the internet of things, a THING may be a human with an implanted heart monitor, a farm animal with a biochip transponder, a vehicle with built-in sensors, DNA analysis equipment, etc. These gadgets gather meaningful data using a variety of current technologies, then autonomously transmit the data to other devices [5].

An IoT ecosystem consists of web-enabled smart devices that use embedded systems such as processors, sensors, and communication gear to gather, transmit, and act upon data from their surroundings [6]. IoT devices share sensed data by connecting to an IoT gateway or other edge device, where the data is either delivered to the cloud or evaluated locally. These gadgets occasionally connect with one another and behave based on the information they receive. People may engage with the gadgets to set them up, give them instructions, or retrieve their data, but they do the bulk of the job without human participation [7]. As common life cycle for an IoT device is as followings:

a. Devices and sensors gather information everywhere. Examples are at home, in the automobile, at work, and at a production facility.
b. Analyzation of the gathered data and obtain information locally. Examples: data visualization, report creation, and data filtering (paring it down).
c. Communication is the transmission of information through networks to a destination. Internet and cloud platforms, Ethernet, and private data centers are examples.
d. Doing some actions based on the information. Send a notice (SMS, email, text) or communicate with another system.

2.1 *The IoT Architecture*

Numerous technologies, including RFIDs and Wireless Sensor Networks (WSNs), may serve as enablers for IoT. The IoT may be considered a vast network consisting of networks of devices and computers linked by a succession of intermediary technologies. RFIDs provide the traceability and addressability of items in real-time. WSNs are the principal instruments used to acquire data from the environment. RFIDs and WSNs are linked to the Internet and have been constituted by embedded intelligence in devices. Therefore, it can attain intelligent control over "things" [8].

In the pharmaceutical sector, IoT facilitates drug manufacturing, storage, packaging, and supply chain management. It aids in enhancing product quality, decreasing mistakes, and boosting pharmaceutical output. It supports pharmaceutical personnel with everyday work schedules and assists them with understanding drugs and active ingredients [9].

After producing the medical product, the good should be transferred to the marketplace. IoT can be employed to monitor completed goods and the supply chain. Internet-enabled embedded computers aid with batch identification. IoT aids in verifying the product's expiration date, and if any batch is recalled from the market, mobile devices make it simpler to find the batch. Thus, it becomes easy to eliminate the same from the supply chain [10]. Moreover, IoT aids clinical trials in real-time management and monitoring of the effects of medications. This real-time data access expedites the detection of adverse responses and associated risks. It contributes to enhancing the quality of the test data [11]. Furthermore, IoT contributes to the improvement of industrial equipment and machinery. The pharma IoT monitoring sensors aid in machine monitoring and provide all pertinent facility data onto a centralized dashboard. By watching, it informs supervisors of aberrant situations and the need for essential maintenance. It also links the automated shutoffs and manages key mechanical situations. Besides, IoT links devices to the internet and network, and linked devices may simply monitor the real-time availability of materials [12].

2.2 IoT Advantages and Disadvantages

Among the benefits of the Internet of Things are the following:

Access to information from anywhere, at any time, on any device; improved communication between connected electronic devices; transferring data packets over a connected network to save time and money; automating tasks to improve the quality of a company's services and reduce the need for human labor are some of the main advantages of using IoT technologies [13]. Besides its benefits, some of the IoT's downsides are the following:

- As the number of linked devices grows and more data is exchanged between them, the likelihood that a hacker may steal sensitive information also rises.
- Eventually, businesses may have to deal with hundreds of thousands, if not millions, of IoT devices, making it difficult to gather and manage the data from all these devices.
- If there is a problem in the system, every linked device is likely to get corrupted.
- Since there is no worldwide standard for IoT interoperability, it is challenging for devices from various manufacturers to connect.

3 Cloud-Based Technology

The word "Cloud computing" refers to a supercomputer which is physically available somewhere and its inner workings are abstracted and connected with the user's computer through the internet. In other words, it is just a large structure loaded with computers (servers). This is a massive data center. Servers serve customers with services such as application execution, data processing, data management and storage, and web hosting [14]. Using the internet, cloud technology enables anyone to access digital resources stored in virtual space. It enables individuals to communicate information and applications regardless of their geographical location. It is a model that enables ubiquitous, convenient, on-demand (pay as you go) network access to a shared pool of configurable computing resources (such as platform, application, processor, and data storage) that can be rapidly provisioned, released, and managed with minimal effort or service provider interaction. Cloud-based data, apps, and platforms are accessible over the internet, and the cloud has eliminated the need for personal computers and other infrastructure [15].

3.1 Advantages of Cloud Computing

Cost savings is the most crucial factor in cloud migration. Moving to the cloud requires no initial expenditure. Typically, companies only need to pay for what they consume. They no longer pay for wasted resources, as would be the case with on-premises servers. Besides, the scalability of cloud computing resources, such as memory, processing power, etc., is very flexible. The company may start small and progressively expand on infrastructure as the firm develops. Cloud computing gives them confidence, regardless of how large their firm develops. They will always be prepared to shoulder the burden. As soon as the demand is decreased, capacity is also decreased to cut costs. A firm may grow storage, computing, and other resources as much or as little as required. Not only can it scale, but it is also more efficient since you can lower capacity when not in use, which is not possible with on-premises servers [16]. Availability is another advantage of using the cloud which implies that the services are accessible 24/7 from everywhere. Even in the event of a malfunction or natural catastrophe, highly available systems can fulfill requests. Cloud computing facilitates catastrophe recovery like never before. Multiple copies of data are kept in various locations/availability zones in the cloud. Therefore, even if one of them fails or a natural catastrophe occurs at the place, the data is secure since it can be restored from another location which has an impact on the reliability of this system. Furthermore, the cloud provides the capability of endless scalability. The services of cloud computing are provided through a network of data centers. These data centers are equipped with robust hardware and software. In addition, they are constantly updated to the newest, most efficient hardware, which results in a high level of performance. One of the benefits of cloud computing is that corporations may store as much data as they like. There is no need to be concerned about disc size. It is like storing data on an infinitely large drive [17].

From the security point of view, numerous cloud service providers now offer a set of rules and sophisticated approaches to boost the overall security and guard against a variety of cyberattacks. As a fact, cloud security is a shared responsibility in which cloud service providers are responsible for the security of the cloud infrastructure, while customers are responsible for the security of their cloud-based applications [18].

3.2 Cloud Technology in Medicine

Currently, there is an evident change in the creation, use, storage, and sharing of healthcare data. In medicine, cloud technology is the ideal alternative to paper-based recording methods, which may be very effective and reliable in several medical sectors. This technology assists medical management systems in enhancing the quality of their health services [19].

In the medicine sector, cloud computing is not only an Internet-based data storage system for EHR and EMR data, but it also creates a real-time network through which many disruptive technologies, including AI, smart embedded devices, and big data, can use for a variety of purposes, including data integrations, personalized medicine, clinical research, and AI-based robotic surgery. These innovative and intelligent technologies may operate under the umbrella of the cloud which is very useful for healthcare companies and medical research institutions through the promotion of telemedicine, mobile health apps, remote monitoring instruments, and other sophisticated healthcare services [20].

3.3 Cloud Technology in the Pharmaceutical Industry

Data and information may come from several sources inside a pharmaceutical or biotech business, including clinical trials, product complaints, medical information call centers, and healthcare provider or patient reports. As a case is handled and additional information is obtained, all information sources inside the organization must remain consistent and harmonious. The manual data harmonization and the cleansing procedure are time-consuming and costly for the majority of businesses. Implementing a cross-development cloud platform with AI-integrated procedures that might drastically cut reconciliation time. It might also increase the data's quality and consistency, decreasing the organization's risk [21]. The storage of data in the cloud, the rate of constant patient interaction, and immediate constant real-time changes as developments occur has the potential to improve quality control, patient compliance, specific winning drug development and release, and simple one-time market entry for multiple market segments [22].

The advantages of cloud technology in the pharmaceutical industry include, but are not limited to, improving the quality of generated data, real-time monitoring, and control of resources, storing product-related information with greater security and transparency, and facilitating better communication across a geographically dispersed portion of the organization. The most significant effect of cloud computing on pharmaceutical businesses is that it reduces their reliance on internal infrastructures and simplifies global operations. With the integration of cloud computing with other new technologies (e.g., AI), it is feasible to identify chemicals with greater affinity and potency for biological targets [23].

As an example, in silico design processes requiring atomic precision, the cloud offers substantial processing capacity. G protein-coupled receptors (GPCRs) are an example of receptors with a complicated structure and little information. The integration of innovative modeling and molecular dynamic approaches with cloud computing leads to the discovery of a novel site or target of activation for a potential molecule or drug, which may not have been detectable using conventional methods. In the preclinical phase of drug development, very significant data from many sources, such as labs and in vivo and in vitro tests, may be gathered and kept in the cloud in a highly secure and efficient way [24, 25].

4 Data Analytics

Data analysis is the act of analyzing, cleaning, manipulating, and modeling data to identify usable information, informing conclusions, and facilitate decision-making. It defines the method for identifying, prioritizing, precisely formulating, and validating the facts required to accomplish a goal. Data collection, which is the systematic collecting and measurement of information on variables of interest, is one of the primary components of data analysis, which allows one to answer stated research questions, test hypotheses, and assess results. The subsequent step is data processing, which happens when acquired data is converted into usable information. Typically, it is carried out by a data scientist or team of data scientists. It is essential that data processing be performed accurately so as not to adversely impact the final product or data output. Data cleansing is the process of correcting or deleting inaccurate, corrupted, improperly formatted, duplicated, or insufficient data from a dataset. Exploratory data analysis is the next level of data analysis. It refers to the crucial process of completing preliminary studies on data to detect trends, identify outliers, test hypotheses, and verify assumptions using summary statistics and graphical representations [26].

It is possible to use mathematical formulae or models (also known as algorithms) or artificial intelligence models and algorithms to the data to find links between the variables; for example, by employing correlation or causation, which results in data products. These items are computer methods that accept data inputs and create outputs, which are then returned to the environment. It may be model- or algorithm-based. Communication of data is the most important aspect of data analysis. It is evident that regular communication is an essential tool for any successful data analysis, and it is essential to recognize that communication is both a tool and a result of data analysis. Upon completion of a data analysis project, you will ultimately want to share your findings and outcomes with an audience [27].

4.1 Big Data Analytics

Nowadays every day, 2.5 quintillion bytes of data are created. 90% of the world's data was produced over the previous two years. The data is too large, moves too quickly, or does not suit the database designs' structures. The data's volume, variety, and complexity need the development of novel architecture, methodologies, algorithms, and analytics to manage it and extract its value. Big data is a collection of data sets that are so huge and complicated that is impossible to handle using typical database management tools and analysis [28].

The significance of big data is in how to use the data that are collected. Data may be retrieved from any source and analyzed to allow cost and time savings, new product creation, and smart decision-making. Combining big data with advanced analytics may have significant effects on corporate strategy to identify the core cause of operational failures, difficulties, and faults in real time. It may also be able to recalculate whole risk portfolios in only minutes and detect fraud before it impacts and endangers the business. Most pharmaceutical firms are swamped with data acquired from production and research facilities. However, the collected data is not well analyzed and used since it is unclear what value it provides or how to harness it for development [29].

Big data analytics is the process of analyzing large amounts of data to identify patterns, unearth trends, and discover undiscovered connections and other helpful information for making quicker and more informed choices. Big data analytics enables firms to harness their data and uncover new possibilities, resulting in wiser business decisions, more efficient operations, more profitability, and satisfied consumers. In the pharmaceutical sector, the introduction of big data analytics simplified and increased the efficiency of complicated business operations. Big data analytics allows the pharmaceutical sector to delve deeply into its data and extract insights from it. To make judgments based on data, many sources, such as social media, sensors, log files, and patient enrollment may provide historical or real-time information for this sector. The goal is to reduce research and development expenses, improve clinical trials, accelerate drug discovery, manage adverse medication reactions, enhance precision medicine, and emphasize collaboration. Future studies on the use of big data in this business should focus on the definition of methods to promote and execute these goals [30, 31].

4.2 Enhancing Data Analytical Capabilities in the Pharmaceutical Industry

Data analytics may have an impact on the following aspects of the pharma industry:

- speeding up bringing a new drug to market by offsetting the growing costs associated with doing so.
- cut costs and accelerate clinical trials by streamlining the whole process and locating areas with high patient availability.
- accelerating illness detection and constructing more effective control groups and clinical trials.
- customizing and developing targeted medications by combining genetic sequencing data, patient medical sensor data, and electronic medical records.
- identifying trends that will aid in the development of more effective and individualized medications by combining genetic sequencing data, patient medical sensor data, and electronic medical records.
- Improving the overall efficiency of the processes by increasing pressure on the operating margins of pharmacies making it imperative.
- making more informed choices by analyzing critical data such as the average ingredient cost per prescription at a granular level to enhance revenue and cut expenses.
- listening to social and search engines to capture data of interest to assist firms in comprehending how their goods are accepted.
- prioritizing efforts and obtaining a competitive advantage by gathering crucial data and identifying new markets through evaluating the efficacy of various marketing channels.
- decreasing expenses by enhancing current operations and procedures and comprehending how machine settings, operator training levels, and inputs of raw materials will impact output quality and optimize and enhance the whole process.
- decreasing the risk of compliance problems by supporting and helping human employees to perform their everyday activities and, if necessary, issuing alarms.

Some of the drawbacks of data analysis in the pharmaceutical industry are as follows:

- Lack of specialized personnel to deal with Big Data analytics.
- Issues related to moving from conventional data processing techniques to real-time data acquisition and analysis methods.
- Exodus of data and data Integration.
- Inconsistencies in electronic health records.

5 Blockchain Technology

Lack of transparency, difficulties in tracing items, a lack of confidence, and the distribution of outdated medicines plague the pharmaceutical sector. Several of these issues might be addressed using blockchain technology. Multiple sectors, including the pharmaceutical business, are undergoing a rapid digital revolution due to blockchain technology. A blockchain is a form of database that stores data in blocks that are then linked together. When the block is filled with fresh data, it is appended to the preceding block. Thus, data are kept in chronological order. In addition, data placed into blocks cannot be modified, making the solution irreversible. In other terms, Blockchain is an interconnected block-based database system. This database can handle any kind of information that has a transaction. The freshly executed transactions are appended to the previous block, and a new block is then generated. These chunks, on the other hand, are temporally interconnected. In this approach, the entering block increases the security of the records by verifying the information in the blocks before it. These blocks are encrypted and have a dispersed structure. Therefore, it becomes very hard for a user to modify the historical records on this network since it also needs to modify the records of other users. The greater the size of the network and the number of unique registrants, the greater its security. This avoids the expenses of verification and auditing and creates a dependable, responsible organization [32].

Blockchain is one of the most adaptable technologies of the modern era. It has several benefits, including improved transparency, increased security and privacy, and accessibility [33]. The pharmaceutical sector understands the promise of blockchain technology and is attempting to utilize it to address current problems. From storing the most crucial data updates to digitizing a supply chain, blockchain has a history of solving problems effectively [34]. Counterfeit (fake) drugs are a perilous and serious problem that authorized makers and authorities have fought for decades. Therefore, producers and even customers were required to devise more effective means of combating this severe danger. The pharmaceutical business used blockchain to aid in the fight against counterfeiters. Using various tagging methods (QR codes, NFC tags, and barcodes), producers might tag manufactured medications and record their information in the blockchain system. Smart contracts on the blockchain are an effective method for registering and storing data without the risk of it being lost or tampered with. The time-stamping function offers greater degrees of transparency. Therefore, it might be used to record clinical studies and show patient consent [35].

There are several motivations to use Blockchain technology in the pharmaceutical sector. These benefits include, but are not limited to:

- Security: information is immutable. All transactions are encrypted and connected to the prior block. Thus, it becomes simpler to track back activities and fraud attempts are eliminated.
- Transparency: the database operation ledge is accessible to the public, making blockchain-powered processes transparent.

- Traceability: It facilitates tracing the origin of the items. It's the greatest way to track a product's travel through the supply chain. When goods change hands, additional transactions will be added to the blockchain.
- Automation: operations may be automated via the use of "smart contracts" that decrease the need for human interaction and dependence on third parties to check compliance with contract requirements. A smart contract is a piece of executable code that contains the real rights and obligations agreed upon by all signees, including the terms and conditions for the payment and delivery of goods and services.

Some of the potential areas of application for this technology are as followings:

- Returned drugs authenticity: periodically, pharmaceutical corporations must deal with returned medications. This occurs when wholesalers overstock their shelves. Therefore, they are required to return the unused product to the makers. Before pharma firms can sell the returned pharmaceuticals on the market, however, they must first detect and isolate the counterfeit drugs included in the returns. By employing decentralized blockchain technology, pharmaceutical makers may quickly record the serial number of packages on a ledger. This indicates that the medicine may be validated from any location.
- Pharmaceutical supply chain improvements: drug supply chain management is one of the primary concerns for society; it is intricate and includes multiple stockholders. The distributed nature of the blockchain database encourages both openness and security. All stockholders participating in the supply chain should adhere to the system's set procedure to evaluate, exchange, and update data and ensure that the information is accurate and delivered on time. Utilizing blockchain technology will assist businesses in reducing the prevalence of fraud on the market. According to research, one in ten medications sold on the market is counterfeit. Blockchain is advantageous in this instance since it can provide swift and secure global transactions/operations, and corporations may use this asset to construct auditable trials and drug sources throughout the supply chain. It indicates that everyone, from producers to customers, might check the drug's source, development process, quality, and delivery circumstances in a secure environment. Moreover, since this form of database is distributed, it increases transparency and safety. All parties participating in the supply chain should adhere to the system's set procedure to evaluate, exchange, and update data and ensure that the information is accurate and delivered on time. This technology will assist firms in reducing the number of market-based scams since the parties will have complete control over the drug development process. Thus, the deployment of distributed systems will benefit mankind on a worldwide scale by improving the performance and quality of businesses and their goods. It would result in improved medical treatments and fewer deadly situations in the future. Additionally, it will enforce trust and openness.
- Data reliability and quality in clinical trials: Blockchain is an impenetrable and secure ecosystem. Therefore, it may boost participants' confidence, and they will be more excited about taking part in tests of novel treatments. The use of

blockchain technology enhances clinical trials since it enables more consent transparency and traceability. The integrity and quality of clinical trial data will also increase with blockchain integration. The decentralized nature enables clinical laboratories to use a transparent, immutable data source where data cannot be altered. As the system is dispersed over several parties and it is almost hard to decode the hash of a block, the system is safe. Even if a portion of the architecture is compromised, the information will not be lost since it is held by other parties. Consequently, it can ensure the confidentiality and accuracy of clinical trial data. In addition, stakeholders may build public registries for accessing clinical trial outcomes.
- Patient data management: using blockchain patients will have the ability to restrict access to their personal sensitive information and share only pertinent account information. In addition, blockchain can function as a database to monitor and store patients' information, allowing organizations across the whole healthcare system to utilize a single identity to access the required data.
- Inventory administration: with supply chain integration, pharmaceutical businesses may better manage their stocks. Additionally, the inventory may be automated to generate triggers based on supply and demand scenarios. For example, if there is an increase in demand, the system may inform the system of demand and request greater medication manufacture.
- Blockchain applications in R&D: the technology may also assist in overcoming R&D cycle obstacles. It has the potential to enhance the drug research and development process since it allows cooperation across industrial, academic, and charitable institutions. For instance, "smart contracts" would shorten the time required to sell a new drug since corporations will not need to invest more resources in the process (banks, lawyers, etc.). The combination of the technology with AI, ML, and NLP will improve data mining and analysis of current literature or patent applications. In addition, the technology's security benefits intellectual property protection, guaranteeing that fresh research data and discoveries cannot be compromised.

6 Telemedicine Innovation

Sitting in close quarters with others at the doctor's office might induce infection. This is particularly dangerous for those with chronic medical conditions or a compromised immune system. The patient may be satisfied with their doctor if they do not have to go to the office, wait for treatment, or get an infection from the hospital. Telemedicine is a health-related service using telecommunications and electronic information technology. It has several applications, such as online patient consultations, remote control, telehealth nursing, and remote physical and psychiatric rehabilitation [36].

Telecommunication is the exchange of information across a considerable distance using electronic and electrical methods. This is a broad phrase that encompasses a variety of information-transmitting technologies and communications infrastructures, including landline phones, mobile devices, cellphones, microwave communications, fiber optics, satellites, radio and television broadcasts, the internet, and telegraphs [37]. Telemedicine is an umbrella word for any medical practice with a remote component. Telemedicine is the delivery of healthcare treatments to patients using electronic communications and software. Although the phrases telemedicine and telehealth are sometimes used interchangeably, they are different. Telehealth is distinct from telemedicine in that it encompasses a larger range of remote healthcare services. Telehealth may refer to distant non-clinical services, such as provider training, administrative meetings, and continuing medical education, in addition to clinical services, while telemedicine refers primarily to remote clinical services [38].

Some of the known telemedicine applications are as followings:

- Subsequent visits: utilizing telemedicine for regular follow-up visits is an efficient way for clinicians and patients, it also enhances the chance of follow-up, hence decreasing missed appointments and enhancing patient outcomes.
- Remote treatment of chronic illness: the rising prevalence of chronic illness poses a significant challenge to our health system. It is a perfect candidate for the use of telemedicine since it makes it simpler and less costly for patients to keep control over their health.
- Preventative care support: weight reduction and quitting smoking are the keys to lowering heart disease and a variety of other diseases. Telemedicine may be a useful tool for connecting clinicians with patients to ensure they get the necessary assistance for success.
- Support for assisted living facilities: telemedicine software has already shown its value in keeping assisted living facility residents out of the hospital. Nighttime and weekend occurrences of difficulties make hospitalization the sole choice, even for less serious issues.

Here are some of the telemedicine advantages:

- For the Patients: less time away from work, no trip expenditures or time, less interference with a kid or elderly care obligations, no exposure to potentially infectious patients.
- For the healthcare providers: increased income, improved office efficiency, a response to the competitive challenge posed by retail health clinics and online-only providers, improved patient follow-up and health outcomes, fewer missed and canceled appointments, private payer reimbursement.

Telemedicine has several uses in pharmaceutical practice. Among them are the following:

- Establishment of pharmaceutical services in a telemedicine critical care unit: the rapid, consistent delivery of critical care and the availability of highly competent critical care doctors, nurses, and pharmacists 24 h a day, seven days a week are significant issues for small hospitals. All hospitals within the system were obliged to adopt order-scanning equipment to enable the remote ICU pharmacy personnel to process orders quickly. The system also may potentially have a computerized physician order input, which leads to orders being received immediately by the pharmacy information system for verification.

- Telemedicine to improve pharmacist services in nursing facilities: the use of telemedicine to help in the provision of care by pharmacists outside of nursing homes has been found to be successful. Since prescription accuracy was validated, telemedicine services offered after hours by pharmacists in three community hospitals lacking 24-h pharmacy facilities increased medication safety. The number of interventions by pharmacists to resolve drug-related issues was five times higher with telemedicine services than without them. By using telemedicine services, hospitals might save an estimated $261,109 annually. After exposure to pharmacist telemedicine services, the nurses' level of comfort with the drug usage procedure increased, as measured by their satisfaction.
- Telepharmacy during the COVID-19 pandemic: Telepharmacy is described as a technique of pharmacy practice in which a pharmacist uses telecommunications technology to supervise pharmacy operations or deliver patient-care services. Telepharmacy operations and services include drug evaluation and monitoring, dispensing, verification of sterile and nonsterile compounding, pharmaceutical therapy management, patient assessment, patient counseling, clinical consultation, outcome assessment, decision support, and drug information. During the pandemic, remote healthcare services have shown advantages for the healthcare system and the improvement of public health. Access to telehealth services allows for greater social distance and decreased infectious risk. In addition, the demand for healthcare facilities was lowered by reducing the number of patients seeking treatment [39].

7 Artificial Intelligence Applications

Artificial intelligence employs computers and technology to simulate the human mind's problem-solving and decision-making skills. The replication of human cognitive processes by machines, primarily computer systems, is known as artificial intelligence. Expert systems, natural language processing, voice recognition, and machine vision are specific AI applications. It has grown into a problem-solving science with enormous applications in business, healthcare, and engineering. Among the objectives of artificial intelligence are computer-assisted learning, thinking, and perception. Weak AI tends to be basic and task-focused, while strong AI performs more complex and human-like activities [40].

AI may be classified as either weak or powerful. Weak AI, also known as narrow AI, is an AI system developed and taught to do a certain job. Weak AI is used by industrial robots and virtual personal assistants such as Apple's Siri. In contrast, strong AI, often known as artificial general intelligence (AGI), refers to software that can mimic the cognitive capabilities of the human brain. A powerful AI system may employ fuzzy logic to apply information from one area to another and independently discover a solution when confronted with a new assignment [41]. There are typically four forms of artificial intelligence:

- Reactive AI: it optimizes outputs based on a set of inputs using a collection of algorithms. AIs that play chess, for instance, are reactive systems that optimize the optimal winning strategy. Reactive AI is often static and incapable of learning or adapting to fresh settings. Thus, equal inputs will result in similar output.
- AI with limited memory: artificial intelligence with limited memory may adapt to prior experience or update itself depending on fresh observations or data. Typically, the quantity of updating is restricted and the length of memory is small. Autonomous cars, for instance, can "read the road" and adapt to unfamiliar conditions, even "learning" from previous encounters.
- AI with a theory of mind: theory-of-mind AI is highly adaptable and has a vast capacity for learning and remembering prior events. These sorts of artificial intelligence include sophisticated chatbots that might pass the Turing Test and convince a human that they are human. While sophisticated and amazing, these AI lack self-awareness.
- Self-aware AI: as the name indicates, self-aware AI is sentient and aware of its own existence. Still, in the realm of science fiction, some scientists think that artificial intelligence will never attain consciousness or "life."

AI describes all sorts of machine intelligence. This includes applications for thinking, language processing, and learning. Machine learning (ML) is a sort of AI that is mainly dependent on statistical approaches and depends on computers' capacity to infer relationships and make predictions rather than human effort. ML is a subfield of AI and computer science that focuses on the use of data and algorithms to emulate how people learn while continuously improving its accuracy.

7.1 AI in the Pharmaceutical Sector

A pharmacist's responsibility has been, for decades, to guarantee that prescriptions are filled with the correct medication at the correct dosage, and in the event of numerous drugs, to ensure that there are no harmful drug-drug interactions. However, the situation has evolved dramatically during the last five years. With the advent of big data and AI, robots are becoming more reliable for physicians, and a growing number of institutions are using robots under human supervision to do formerly human-only tasks [42].

Pharmaceutical firms have access to a big number of substances that could tackle a large number of particular ailments. Nonetheless, the firms have no means of identifying themselves as such. Drug research and manufacture is a difficult endeavor that may cost a pharmaceutical business up to $2.6 billion and take 12–14 years to complete. Here, AI becomes advantageous for pharmaceutical corporations. AI decreases the time required for medication development, which in turn reduces the related expenses, increases the returns on investment, and may even result in a drop in cost for the end user [43].

To assure the safety of medication users, AI is increasingly being employed in the pharmaceutical industry to determine the optimal doses of drug consumption. Not only does it aid in patient monitoring during clinical trials, but it also advises the appropriate dose at regular intervals. Most artificial intelligence solutions utilized in healthcare today are based on data science algorithms developed by humans. This sort of artificial intelligence employs multivariate data analytics supplemented by prior experience data. It might, for instance, integrate population-based treatment results with an individual patient's clinical data and medical history to generate therapy choices and prescribe medication combinations. ML also can make judgments much more quickly and correctly. Using data-driven methods, machine learning enables software to anticipate outcomes with high accuracy without the need for explicit programming. Deep learning, which is also based on neural networks, combines distinct layers of computations with integrated data. Deep learning has enormous diagnostic potential since it can properly interpret visuals (such as photographs of skin disorders or radiological scans) in conjunction with pathology data and past treatment results. AI might identify the ideal candidates for clinical trials using social media and medical visits, hence increasing the success rates of new treatments, and reducing operating expenses.

For manufacturing process improvement, AI can perform quality control, reduce design time, reduce material waste, enhance manufacturing reuse, and predict maintenance, among other tasks. AI may be used in a variety of ways to increase industrial efficiency, resulting in quicker output and less waste [44]. From developing new compounds to finding novel biological targets, AI is contributing to drug target identification and validation; target-based, phenotypic, and multi-target drug discovery; drug repurposing; and biomarker identification. Perhaps the most advanced use of AI to date is in algorithms intended to read, organize, and understand vast amounts of textual material. This may be a significant time saving for researchers in the life sciences business, as it provides a more efficient approach to review the larger-than-usual volumes of data from the expanding number of research papers to confirm or reject ideas [45]. In addition, many clinical trials continue to depend on paper diaries in which patients record when they took a medicine, what other prescriptions they were taking, and any adverse responses. AI can gather and analyze anything from handwritten notes and test results to environmental elements and imaging images. The advantages of using AI in this manner include speedier research and data cross-referencing, as well as merging and extracting data into analysis-friendly forms. Clinical data extraction and interpretation from medical records are in great demand in the medical and pharmaceutical sectors, and AI may be the solution [20].

In addition to assisting to make sense of clinical trial data, another use of AI in the pharmaceutical sector is the recruitment of study participants. The AI is capable of extracting tens of thousands of clinical data points to generate a multidimensional profile. Then, physicians and researchers may use these profiles to discover candidates for clinical trials. As this is being done by a machine, the procedure is far faster and more accurate. Using sophisticated predictive analytics, AI can evaluate genetic data to determine the best sample size and the right patient group for a clinical study. Some AI technology can interpret unstructured data, such as doctor's notes and intake paperwork, as well as free-form language entered by patients into clinical trial software [46].

AI is used to construct a virtual assistant that supports clinical project management and enables researchers to enhance clinical trial performance and management. This AI-based technology analyses patterns and large data to save the correct and online patient information, enabling researchers to anticipate the patient's illness development and treatment response, particularly for clinical trial stratification and diagnostics [47].

Among the most time- and money-saving uses of AI is the capacity to match medication treatments with specific patients, therefore lowering the amount of trial-and-error effort previously required [48]. ML models can predict a patient's response to potential drug treatments by inferring potential relationships between factors that may be influencing the results, such as the body's ability to absorb the compounds, their distribution within the body, and a person's metabolism [49].

Biomarker development is crucial not just in the context of medical diagnostics, but also in the context of drug discovery and development. For instance, predictive biomarkers are used to detect probable responders to a molecularly targeted medication prior to human testing. In this approach, AI employs biomarker models "trained" using massive datasets [50].

Pharmaceutical firms have a significant challenge in ensuring volunteer participants in clinical trials adhere to a drug research procedure. If participants in a drug trial violate the trial's guidelines, they must be removed from the experiment or risk tainting the trial's findings. One of the most critical aspects of a successful drug trial is ensuring that participants take the specified doses and timings of the medicine under study. It is crucial to have a method for ensuring medication adherence for this reason. Through remote monitoring and algorithms for analyzing test results, AI can separate the excellent from the poor. AI smartphone applications ("apps") may be suitable as instruments for measuring and promoting medication adherence to monitoring medication adherence in various individuals and treatment procedures [51, 52]. AI technology can also be used to analyze large amounts of data to identify patterns, which is the primary reason why this technology is being utilized more frequently for disease mapping and identifying demographics of people in order to investigate the health conditions of societies and for public health surveillance [53].

In the pharmaceutical sector, AI may be used to develop and target advertising materials and find the most effective routes for contacting prospective clients. AI may also be used to monitor and predict client preferences and behavior in this industry. This data may then be utilized to personalize communications and offers, boosting their likelihood of acceptance. Pharmaceutical firms may enhance their business potential, marketing tactics, sales productivity, and customer service by embracing AI [54].

In the pharmaceutical industry, AI excels at automating tedious processes such as data input and lead creation. This system may offer sales representative divisions with real-time feedback. When introducing new goods, these essential real-time feedbacks allow them to modify the new strategy and avoid development job delays [55].

AI can provide clients and consumers with individualized suggestions or proactively addressing their wants and problems. It may also be used to automate customer care operations such as replying to commonly requested queries. Pharmaceutical businesses already generate vast quantities of data. AI can filter through information such as trends and insights so that they are not lost. AI can perform relatively ordinary jobs exceptionally quickly; provide corporations with methods to generate new goods at cheaper prices; produce new medications considerably more quickly than in the past; reduce the number of new drug failures; and perform tasks without human mistakes [56].

8 Wearable Medical Device Technology

Medical wearables are devices that may be used to monitor health parameters such as heart rate, blood pressure, oxygen saturation, and several other vital indicators, and can offer crucial information for the diagnosis and treatment processes to healthcare experts. These devices are suggested for illness detection, treatment, and prevention [57]. The use of wearable technology provides the following evident benefits:

- Reduced healthcare costs by detecting and treating possible health concerns early; numerous medical wearable devices are able to detect physical activity and mental actions, which might motivate healthy behavior.
- Improved patient care: medical wearables may provide healthcare providers with useful information. Examples of medical wearable technologies include smart garments that monitor health status and vital signs, smart shoes that assess running style, smart glasses that provide hands-free internet access, and smartwatches that measure a variety of medical parameters.

Monitoring chronic illnesses such as oncological disease, respiratory problems, and cardiovascular disease might rely heavily on wearable devices. In addition to basic healthcare monitoring, wearables are now being utilized to aid in the treatment of chronic conditions such as asthma and diabetes. The most prevalent chronic illness among children and the elderly is asthma. Wearables equipped with sensors that can monitor the patient's breathing and oxygen levels, as well as environmental variables like pollen count, are used to prevent or reduce asthma episodes. In addition, certain wearable items can now test glucose levels by placing sensors just under the skin, allowing diabetic patients to monitor and regulate their blood sugar levels in real-time. However, this technology has its own inherent dangers. The hazards of technological faults and omissions, data security problems, high cost, and physical damage are examples [58, 59].

9 Cyber-Physical Systems

A cyber-physical system (CPS) or intelligent system is a computer system in which computer-based algorithms operate or monitor a mechanism. These systems are characterized as the evolving collection of distributed cyber systems that regulate the linked physical systems networked by control loops based on a set of rules. It incorporates a network of sensors that gather data that is examined and processed by a control processing unit, which then sends instructions to control actuators through communication devices [60].

A CPS includes both a physical and a cyber system. Integration of physical processing, sensing, computing, communication, and control leads to it. The cyber system consists of communication networks, computer facilities, and command and control centers. Typically, physical processes are seen as a plant managed by a cyber system [61]. As for other components, their functions are as follows:

- Sensors are employed for data collecting in real-time.
- Actuator: Actuators execute control orders to accomplish the intended physical operations.
- The computing and control center is responsible for receiving sensor data. The control center ensures that physical operations are carried out properly by assessing the collected data and making the appropriate control choices.

The paradigm of cyber-physical systems influences the healthcare system via the integration of distributed cyber systems that monitor, and control based on integrated appropriate functions of physically interconnected systems to provide a personalized medical assessment or even to predict medical events. Consequently, the CPS healthcare network is a dispersed system in which all devices are networked, exchanging data, and sharing information over a competent communication network. Monitoring the patient's parameters, aiding in remote surgical procedures, monitor and giving help to patients in any condition, as well as remote diagnostic and remote medical care support, are some of the CPS's applications in the health network [62].

CPS in healthcare systems influences; the health of the general population by analyzing data to find adaptive treatments; anticipating and predicting the health needs of patients; producing personalized medical results; matching organs for transplant; and increasing access to health care [63].

Event-based data gathered from medical records and real-time data are examples of health data that might be utilized in CPSs. The data may be acquired in a variety of methods, including in a dispersed form stored locally by sensors and devices or through a data warehouse. Typically, the CPSs' communication layer is responsible for network connection and data transfer. It indicates that CPS is designed as an interoperable system that electronically links health systems to exchange information collectively. The science of CPS has the potential to influence technology in a wide range of industries and organizations, which will change medicine, the pharmaceutical industry, and the process of health by providing doctors with predictions about patients' illnesses, risks, epidemics, and problems in the general public's health before they occur, as well as early-stage remedies or treatments, drug discovery, and personalized healthcare packages and services [64, 65].

CPS systems are susceptible to different cyber and/or physical security risks, assaults, and obstacles despite their considerable benefits. This is a result of its diverse character, dependence on private and sensitive data, and widespread implementation. Consequently, purposeful, or inadvertent exposures of these systems might have devastating consequences, making it imperative to implement stringent security measures. However, this might result in an untenable network burden, particularly in terms of latency. In addition, frequent software, application, and operating system upgrades should be used to reduce zero-day vulnerabilities. In addition, the security problem in the pharmaceutical business remains an unmitigated risk or weakness in this technology. This risk factor may lead to identity theft and fraud via the theft of patient records, the loss of medical information, disruptive attacks, and the shutdown of hospital information systems [66].

10 Pharma 4.0

Pharma 4.0 is the application of Industry 4.0 to the pharmaceutical industry. Industry 4.0 is the fourth revolution in the industrial sector. It captures the changes that the CPS has brought about in the pharmaceutical industry [67].

In Industry 4.0, machines are able to interact with one another, evaluate data, and notify humans as required. Using IoT, the machines will communicate with one another wirelessly. During this time, the system will operate at a lesser cost and produce more swiftly. They are the digital twins responsible for managing machinery, analyzing data, and making other production choices. Pharma 4.0, in its most fundamental sense, is the incorporation of cutting-edge digital components and enabling technologies into the current Pharmaceutical Quality System. Throughout the whole lifespan of pharmaceutical goods, more opportunities for process improvements will be available. There will be a higher degree of interconnectedness and transparency, allowing for tighter control over operations and quality, as well as quicker decision-making. Due to the increasing vulnerability of networked automated systems, moving to Pharma 4.0 will need increased levels of security [68].

Pharma 4.0 is a digital framework for integrating digital industrial technologies into pharmaceutical manufacturing. Similar to Industry 4.0, Pharma 4.0 aims to increase production and responsiveness while decreasing waste and costs [69, 70].

In 2017, the International Society for Pharmaceutical Engineering (ISPE) coined the phrase "pharma 4.0." The organization characterized Pharma 4.0 as a "holistic operating model for pharmaceutical factories and supply chains of the future based on Industry 4.0 capabilities, digital maturity, and data integrity." Pharma 4.0 refers to the proactive, continual interchange of information about pharmaceutical processes and products, from drug development through patient care. With Pharma 4.0, pharmaceutical businesses may increase their transparency and quality while decreasing their total expenses [71].

Some of the advantages of shifting towards Pharma 4.0 are as followings:

- High degree of process awareness, scientific and risk-based approach, critical thinking, and transparent controls and activities.
- New workforce—minimizing human interference.
- Interdisciplinarity: controlling, enhancing, and optimizing processes in real-time from the patient's perspective.
- Increased quality: by digitizing pharma production processes and collecting data, organizations may more readily discover process abnormalities and implement real-time corrections. This may lead to improved batch quality and consistency.
- Reduced expenses: by automating manual activities, eliminating downtime, and reducing waste, Pharma 4.0 may assist businesses in lowering costs. In addition, the improved quality that comes with Pharma 4.0 may help businesses minimize the expenses associated with product recalls and other quality concerns.
- Improved compliance: it gives more insight into pharmaceutical production processes, and it may assist businesses in ensuring regulatory compliance. Automated data collecting and analysis may also speed up the identification and correction of deviations from normal operating procedures.
- Greater flexibility: it enables enterprises to rearrange their production processes, hence increasing their flexibility. This might allow for changes in product demand or the rapid introduction of new items.
- Enhanced safety: the capacity to trace items across the supply chain enables businesses to enhance product quality and safety.

Additionally, the automation of pharmaceutical production processes may limit worker exposure to dangerous substances.

Conclusion

In conclusion, digital technologies have the potential to revolutionize the pharmaceutical industry by improving efficiency, accuracy, and cost-effectiveness. From drug development and clinical trials to supply chain management and patient care, digital technologies are transforming the way the pharmaceutical industry operates. Despite the challenges that must be overcome, including data privacy and security, the benefits of embracing digital technologies in the pharmaceutical industry are clear. In the coming years, it is likely that the use of digital technologies will continue to grow, and that the pharmaceutical industry will increasingly look to these technologies to improve the development and delivery of new drugs, and to enhance patient care. To fully realize the benefits of digital technologies in the pharmaceutical industry, it is essential that stakeholders work together to overcome the challenges that lie ahead and to promote the adoption of these technologies. Ultimately, the successful integration of digital technologies into the pharmaceutical industry will result in a more efficient, effective, and cost-effective healthcare system that better serves the needs of patients and stakeholders alike.

References

1. Paul SM, Mytelka DS, Dunwiddie CT, Persinger CC, Munos BH, Lindborg SR, Schacht AL (2010) How to improve R&D productivity: the pharmaceutical industry's grand challenge. Nat Rev Drug Discov 9(3):203–214
2. Deore AB, Dhumane JR, Wagh R, Sonawane R (2019) The stages of drug discovery and development process. Asian J Pharm Res Dev 7(6):62–67
3. Schweitzer SO, Lu ZJ (2018) Pharmaceutical economics and policy: perspectives, promises, and problems. Oxford University Press
4. Laurenza E, Quintano M, Schiavone F, Vrontis D (2018) The effect of digital technologies adoption in healthcare industry: a case based analysis. Bus Process Manag J 24(5):1124–1144
5. Tiwary A, Mahato M, Chidar A, Chandrol MK, Shrivastava M, Tripathi M (2018) Internet of Things (IoT): research, architectures and applications. Int J Future Revolution Comput Sci Commun Eng 4(3):23–27
6. An IoT ecosystem consists of web-enabled smart devices that use embedded systems such as processors, sensors, and communication gear to gather, transmit, and act upon data from their surroundings
7. Tawalbeh LA, Muheidat F, Tawalbeh M, Quwaider M, Abd El-Latif AA (2022) Edge enabled IoT system model for secure healthcare. Measurement 191:110792
8. Barenji RV, Barenji AV, Hashemipour M (2014) A multi-agent RFID-enabled distributed control system for a flexible manufacturing shop. Int J Adv Manuf Technol 71:1773–1791
9. Sharma A, Kaur J, Singh I (2020) Internet of things (IoT) in pharmaceutical manufacturing, warehousing, and supply chain management. SN Comput Sci 1:1–10
10. Ahmadi V, Benjelloun S, El Kik M, Sharma T, Chi H, Zhou W (2020) Drug governance: IoT-based blockchain implementation in the pharmaceutical supply chain. In: 2020 sixth international conference on mobile and secure services (MobiSecServ), pp 1–8. IEEE
11. Badulescu Y, Tiwari MK, Cheikhrouhou N (2022) MCDM approach to select IoT devices for the reverse logistics process in the Clinical Trials supply chain. IFAC-PapersOnLine 55(10):43–48
12. Singh M, Sachan S, Singh A, Singh KK (2020) Internet of Things in pharma industry: possibilities and challenges. In: Emergence of pharmaceutical industry growth with industrial IoT approach, pp 195–216
13. Sharma DK, Kaur G, Sharma M (2021) An industrial internet of things approach for pharmaceutical industry growth. In: Further advances in internet of things in biomedical and cyber physical systems, pp 289–309
14. Rashid A, Chaturvedi A (2019) Cloud computing characteristics and services: a brief review. Int J Comput Sci Eng 7(2):421–426
15. Attaran M, Woods J (2019) Cloud computing technology: improving small business performance using the Internet. J Small Bus Entrep 31(6):495–519
16. Namasudra S (2021) Data access control in the cloud computing environment for bioinformatics. Int J Appl Res Bioinform (IJARB) 11(1):40–50
17. Cai F, Zhu N, He J, Mu P, Li W, Yu Y (2019) Survey of access control models and technologies for cloud computing. Clust Comput 22:6111–6122
18. Alouffi B, Hasnain M, Alharbi A, Alosaimi W, Alyami H, Ayaz M (2021) A systematic literature review on cloud computing security: threats and mitigation strategies. IEEE Access 9:57792–57807
19. Dang LM, Piran MJ, Han D, Min K, Moon H (2019) A survey on internet of things and cloud computing for healthcare. Electronics 8(7):768
20. Hariry RE, Barenji RV, Paradkar A (2020) From Industry 4.0 to Pharma 4.0. In: Handbook of smart materials, technologies, and devices: applications of industry 4.0, pp 1–22
21. Sun L, Jiang X, Ren H, Guo Y (2020) Edge-cloud computing and artificial intelligence in internet of medical things: architecture, technology and application. IEEE Access 8:101079–101092

22. Priporas CV, Vellore-Nagarajan D (2022) New-normal market entry mode for pharmaceuticals: an internet of things (IoT) market entry framework stemming from COVID-19. Int Market Rev (ahead-of-print)
23. Singh S, Sundram S, Malviya R, Sekar M (2023) Drug development using cloud application. In: Bioinformatics tools and big data analytics for patient care, pp. 159–180. Chapman and Hall/CRC
24. Noonan T, Denzinger K, Talagayev V, Chen Y, Puls K, Wolf CA, Wolber G, et al (2022) Mind the gap—deciphering GPCR pharmacology using 3D pharmacophores and artificial intelligence. Pharmaceuticals 15(11):1304
25. Gateau T, Ayaz H, Dehais F (2018) In silico vs. over the clouds: on-the-fly mental state estimation of aircraft pilots, using a functional near infrared spectroscopy based passive-BCI. Front Hum Neurosci 12:187
26. Hulsen T, Jamuar SS, Moody AR, Karnes JH, Varga O, Hedensted S, McKinney EF, et al (2019) From big data to precision medicine. Front Med 34
27. Agbehadji IE, Awuzie BO, Ngowi AB, Millham RC (2020) Review of big data analytics, artificial intelligence and nature-inspired computing models towards accurate detection of COVID-19 pandemic cases and contact tracing. Int J Environ Res Public Health 17(15):5330
28. Hariri RH, Fredericks EM, Bowers KM (2019) Uncertainty in big data analytics: survey, opportunities, and challenges. J Big Data 6(1):1–16
29. Hamilton S (2021) Real-time big data analytics, sustainable Industry 4.0 wireless networks, and Internet of Things-based decision support systems in cyber-physical smart manufacturing. Econ Manag Financ Markets 16(2):84–94
30. Shahbaz M, Gao C, Zhai L, Shahzad F, Luqman A, Zahid R (2021) Impact of big data analytics on sales performance in pharmaceutical organizations: the role of customer relationship management capabilities. PLoS ONE 16(4):e0250229
31. Hariry RE, Barenji RV, Paradkar A (2022) Towards Pharma 4.0 in clinical trials: a future-orientated perspective. Drug Discovery Today 27(1):315–325
32. Vatankhah Barenji R (2022) A blockchain technology based trust system for cloud manufacturing. J Intell Manuf 33(5):1451–1465
33. Saadati Z, Zeki CP, Vatankhah Barenji R (2021) On the development of blockchain-based learning management system as a metacognitive tool to support self-regulation learning in online higher education. Interact Learn Environ 1–24
34. Ghadge A, Bourlakis M, Kamble S, Seuring S (2022). Blockchain implementation in pharmaceutical supply chains: a review and conceptual framework. Int J Prod Res 1–19
35. Uddin M (2021) Blockchain Medledger: Hyperledger fabric enabled drug traceability system for counterfeit drugs in pharmaceutical industry. Int J Pharm 597:120235
36. Haleem A, Javaid M, Singh RP, Suman R (2021) Telemedicine for healthcare: capabilities, features, barriers, and applications. Sens Int 2:100117
37. Li JPO, Liu H, Ting DS, Jeon S, Chan RP, Kim JE, Ting DS, et al (2021) Digital technology, tele-medicine and artificial intelligence in ophthalmology: a global perspective. Prog Retinal Eye Res 82:100900
38. Kaplan B (2020) Revisiting health information technology ethical, legal, and social issues and evaluation: telehealth/telemedicine and COVID-19. Int J Med Inform 143:104239
39. Unni EJ, Patel K, Beazer IR, Hung M (2021) Telepharmacy during COVID-19: a scoping review. Pharmacy 9(4):183
40. Vrontis D, Christofi M, Pereira V, Tarba S, Makrides A, Trichina E (2022) Artificial intelligence, robotics, advanced technologies and human resource management: a systematic review. Int J Hum Resour Manage 33(6):1237–1266
41. Fjelland R (2020) Why general artificial intelligence will not be realized. Humanit Soc Sci Commun 7(1):1–9
42. Ahuja AS (2019) The impact of artificial intelligence in medicine on the future role of the physician. Peer J 7:e7702
43. Paul D, Sanap G, Shenoy S, Kalyane D, Kalia K, Tekade RK (2021) Artificial intelligence in drug discovery and development. Drug Discovery Today 26(1):80

44. Barenji RV, Akdag Y, Yet B, Oner L (2019) Cyber-physical-based PAT (CPbPAT) framework for Pharma 4.0. Int J Pharm 567:118445
45. Chan HS, Shan H, Dahoun T, Vogel H, Yuan S (2019) Advancing drug discovery via artificial intelligence. Trends Pharmacol Sci 40(8):592–604
46. Kaul V, Enslin S, Gross SA (2020) History of artificial intelligence in medicine. Gastrointest Endosc 92(4):807–812
47. Gamberini L, Pluchino P, Porto L, Zordan F, Monterosso A, Palanivel K Fontanari A (2021) Patchai: an e-health application powered by an AI virtual assistant to support patients in their clinical trials. In: HCI international 2021-posters: 23rd HCI international conference, HCII 2021, virtual event, July 24–29, 2021, proceedings, Part III 23, pp 315–322. Springer International Publishing
48. Noorbakhsh-Sabet N, Zand R, Zhang Y, Abedi V (2019) Artificial intelligence transforms the future of health care. Am J Med 132(7):795–801
49. Adam G, Rampášek L, Safikhani Z, Smirnov P, Haibe-Kains B, Goldenberg A (2020) Machine learning approaches to drug response prediction: challenges and recent progress. NPJ Precis Oncol 4(1):19
50. Zhavoronkov A (2018) Artificial intelligence for drug discovery, biomarker development, and generation of novel chemistry. Mol Pharm 15(10):4311–4313
51. Woo M (2019) An AI boost for clinical trials. Nature 573(7775):S100–S100
52. Harrer S, Shah P, Antony B, Hu J (2019) Artificial intelligence for clinical trial design. Trends Pharmacol Sci 40(8):577–591
53. Bansal A, Padappayil RP, Garg C, Singal A, Gupta M, Klein A (2020) Utility of artificial intelligence amidst the COVID 19 pandemic: a review. J Med Syst 44:1–6
54. Racine E, Boehlen W, Sample M (2019) Healthcare uses of artificial intelligence: challenges and opportunities for growth. In: Healthcare management forum, vol 32, no 5, pp 272–275). SAGE Publications, CA, Los Angeles
55. Dash R, McMurtrey M, Rebman C, Kar UK (2019) Application of artificial intelligence in automation of supply chain management. J Strateg Innov Sustain 14(3):43–53
56. Kulkov I (2021) The role of artificial intelligence in business transformation: a case of pharmaceutical companies. Technol Soc 66:101629
57. Mück JE, Ünal B, Butt H, Yetisen AK (2019) Market and patent analyses of wearables in medicine. Trends Biotechnol 37(6):563–566
58. Lu L, Zhang J, Xie Y, Gao F, Xu S, Wu X, Ye Z (2020) Wearable health devices in health care: narrative systematic review. JMIR mHealth uHealth 8(11):e18907
59. DeVore AD, Wosik J, Hernandez AF (2019) The future of wearables in heart failure patients. JACC Heart Fail 7(11):922–932
60. Saadati Z, Barenji RV (2022) Toward Industry 5.0: cognitive cyber-physical system. In: Industry 4.0: technologies, applications, and challenges, pp 257–268. Springer Nature Singapore, Singapore
61. Yaacoub JPA, Salman O, Noura HN, Kaaniche N, Chehab A, Malli M (2020) Cyber-physical systems security: limitations, issues and future trends. Microprocess Microsyst 77:103201
62. Qiu H, Qiu M, Liu M, Memmi G (2020) Secure health data sharing for medical cyber-physical systems for the healthcare 4.0. IEEE J Biomed Health Inform 24(9):2499–2505
63. Sharma A, Rathee G, Kumar R, Saini H, Varadarajan V, Nam Y, Chilamkurti N (2019) A secure, energy-and sla-efficient (sese) e-healthcare framework for quickest data transmission using cyber-physical system. Sensors 19(9):2119
64. Wang S, Wang H, Li J, Wang H, Chaudhry J, Alazab M, Song H (2020) A fast CP-ABE system for cyber-physical security and privacy in mobile healthcare network. IEEE Trans Ind Appl 56(4):4467–4477
65. Hussain I, Park SJ (2021) Big-ECG: cardiographic predictive cyber-physical system for stroke management. IEEE Access 9:123146–123164
66. Vangipuram SL, Mohanty SP, Kougianos E (2021) CoviChain: a blockchain based framework for nonrepudiable contact tracing in healthcare cyber-physical systems during pandemic outbreaks. SN Comput Sci 2:1–16

67. Steinwandter V, Borchert D, Herwig C (2019) Data science tools and applications on the way to Pharma 4.0. Drug Discovery Today 24(9):1795–1805
68. Arden NS, Fisher AC, Tyner K, Lawrence XY, Lee SL, Kopcha M (2021) Industry 4.0 for pharmaceutical manufacturing: preparing for the smart factories of the future. Int J Pharm 602:120554
69. Hariry RE, Barenji RV, Azizi A (2022) Toward Pharma 4.0 in drug discovery. In: Industry 4.0: technologies, applications, and challenges, pp 221–238. Springer Nature Singapore, Singapore
70. Reinhardt IC, Oliveira JC, Ring DT (2020) Current perspectives on the development of industry 4.0 in the pharmaceutical sector. J Ind Inf Integr 18:100131
71. Kumar SH, Talasila D, Gowrav MP, Gangadharappa HV (2020) Adaptations of Pharma 4.0 from Industry 4.0. Drug Invention Today 14(3)

Mathematical Modeling and Hybrid Adaptive-Fuzzy Control of HIV/AIDS Infection

A. Khashayar, A. Izadi, M. Naderi Soorki, and M. Nikbakht

Abstract The Human Immunodeficiency Virus (HIV) is a kind of low duplicating retrovirus called AIDS. In this paper, modeling and hybrid Adaptive-Fuzzy Control of AIDS mathematical model are investigated. By virus infiltration into the body, lymphocyte cells which have a fundamental role in viral elimination will be destroyed and body immune system will be smoothly collapsed which caused very serious defective in exposure to other possible infectious diseases. Nowadays HIV is controlled by using a combination of multiple anti-viral drugs and would be treated with several practical medications. Inhibitor drugs used for treatment divide into two different categories of reverse, transcriptase and protease fields. So in this paper, both of them are considered. To this end, first, a new mathematical modeling of HIV system is presented which provides the best dosages of treatment medicines. Then, using an adaptive self-tuning regulator, the disease is treated by eliminating the viruses around zero. For confirmation to get the best response, a hybrid type of controller in which the fuzzy logic is used as a supportive controller for removing viruses along with an adaptive controller is proposed and its stability is shown using Lyapunov theory. In this controller the upper bound of disturbance term of HIV/AIDS model along with the input control parameters are estimated using some stable adaptive laws. The simulation results reveal that the hybrid fuzzy-adaptive controller has a significant impact on reducing the amount of virus and thus treating AIDS in comparison with other references.

A. Khashayar
South Branch, Department of Electrical Engineering, Azad University of Tehran, Tehran, Iran

A. Izadi
Department of Electrical Engineering, Central Tehran Branch,
Islamic Azad University, Tehran, Iran
e-mail: Ali.Izadi.Eng@iauctb.ac.ir

M. N. Soorki (✉)
Electrical Engineering Department, Sharif University of Technology, Tehran, Iran
e-mail: mojtabanaderi@aut.ac.ir

M. Nikbakht
Department of Mathematics, Payam Noor University, Tehran, Iran
e-mail: M_Nikbakht@pnu.ac.ir

© The Author(s), under exclusive license to Springer Nature Singapore Pte Ltd. 2023
A. Azizi (ed.), *Control Engineering in Mechatronics*, Emerging Trends in Mechatronics,
https://doi.org/10.1007/978-981-16-7775-5_5

Keywords Human immunodeficiency virus (HIV) · Adaptive control · Self-tuning regulator · Control system · AIDS modeling · Nonlinear systems · Fuzzy logic

1 Introduction

In recent years because of the high prevalence of HIV in the world, the control engineers' attention is drawn to the disease and some important control methods have been applied to the models of HIV optimistically, having multivariable nonlinear unstable non-minimum phase dynamics for important disease. HIV/AIDS model puts it in gracious systems category to test the performance and efficiency of the medicines and drugs. Precision studies on it try to analyze the composition and clinical outcomes which have been applied to it shall be stating that disease in the early stages of infection can be controlled as well as possible.

Thus, by using nonlinear analytic control, the number of inquiries to AIDS treatment strategy would be increased with the principles of controllability partially. Some using methods would be mentioned briefly: In [1], referred to a delayed intracellular AIDS model which used the optimal control method to reduce infected cells in the body. The number of cells exposed to contamination could be considered and analyzed with Lyapunov control theory of stability in [2]. In [3] also the principle of Lassalle's instability and Lyapunov control theory has been successfully done to get a reasonable result to control viral infections. In [4] the effect of two essential medicines was revealed by using the impact controller. In [5] the problem of disease evolution was formed by multidimensional optimal control method from a multi-purpose perspective deeply. A nonlinear feedback used to control HIV model even though it took a long time to find a suitable cure, in [6]. Using Roth-Hurwitz the amount of drugs used for viral cure was investigated while the effect of medications on the system model was demonstrated by Ljung [7]. An impact control strategy based on a detailed linearized model of HIV proofing in [8] and given combined with defect frequency impact loss. In order to achieve control aims, two control loops were implemented, inner one for feedback linearization of system dynamics of HIV and one outer for optimal control was depicted by Rivadeneira and Moog [9] carefully. In [10], the predictive control method was proposed for HIV model revealing the robustness of parameter changes. In [11], an optimal impact controller was introduced to HIV system, so a cost function was used to optimize the parameters by Zarei et al. [12], also it used an optimal fuzzy control to system model which applied the genetic algorithm for fuzzy rule design.

Modeling and control of AIDS and HIV infection have been also the focus of control engineers who are working on disease modeling in the last two or three years. This attention is both about modeling and also applying different control methods. In the following, we review the articles of the last two or three years in the field of HIV/AIDS Infection modeling and control.

In [13] a combination dynamic model of HIV/AIDS and Pneumonia is introduced in which 6 equilibrium points isbased on HIV/AIDS or pneumonia disease free situations. The analysis in this paper is based on treatment and vaccination. In [14] two different schemes are presented both numerically and graphically which describes dynamic model of HIV virus in HIV infection. These two methods for mathematical modeling of HIV infection are finally compared to some traditional methods of HIV modeling such as Genetic Algorithm or Laplace domain modeling in the sense of accuracy and efficiency [14].

Recently a mathematical model of HIV/AIDS, which includes important segments such as susceptible and uninformed persons, undiagnosed HIV infections, HIV patients with and without AIDS symptoms, and those treated for this disease developed in [15]. In this modeling, the number of effective reproduction, balance and nature of their stability are formulated and a case study is also presented in Ethiopia. In [16], a new delay-based nonlinear model for the HIV/AIDS disease is presented. Then, the equilibrium points of the model are obtained and its stability is checked with some simple methods such as the Routh–Hurwitz criterion. Finally, it has been shown that a small amount of delay causes disease control in the population and a large amount of delay causes an epidemic of this disease. The rate of incidence and epidemic of AIDS with HIV infection is investigated in [17]. In this case, due to the long and variable period of time between HIV infection and the development of AIDS, epidemiological mathematical modeling of this disease is presented. For this purpose, mathematical structures based on different methods such as back calculations have been used.

Among other references that have addressed the modeling of HIV/AIDS disease from different aspects, we can refer to [18–21]. A model based on latently infected cells in [18], a host HIV/HTLV-I co-infection modeling in [19], model of HIV/AIDS Stigma effect on HIV Infection Dynamics in [20]. Fractional order calculus is a recent method for modeling systems more accurately [21, 22]. In this method, fractional orders are used instead of traditional integer orders and research papers have been devoted to the development of investigation tools for fractional order systems [23–25]. About HIV/AIDS modeling, an epidemic model for HIV/Aids based on fractional order calculations and Mittag-Leffler function in [26] is the newest reference about modeling of HIV/AIDS infection.

After modeling, the second most important issue in dealing with HIV/AIDS infection is its control or stabilization. In last three years, various control methods by providing different treatments have been proposed by researchers in order to stabilize and control this disease. Optimal control is one of the methods that have been used more than others to control this disease [27–29]. In [27] optic control based on Pontryagins-Minimum Principle. In proposed, in [28] a new optimal control method named optimal control power series technique is used and in [29] based on Runge–Kutta scheme, the problem of optimal control using Pontryagin's Maximum Principle is solved in order to find the best strategy of control input combinations. In this paper, a sensitivity analysis of the HIV/AIDS-resistant model is also presented [29].

Robust control is another method for stabilizing HIV/AIDS models in the presence of uncertainties. In [30], a drug treatment is used for controlling HIV infection model in the presence of modeling uncertainties. In this paper with the assumption that measuring all the states in the clinical situation is not possible, a nonlinear disturbance observer is also used in the control method. Another recent reference in this field is [31], where a robust super-twisting sliding mode controller is developed along with an observer which uses output information for stabilizing the nonlinear dynamical equation of HIV infection treated by drugs.

On the other hand, fuzzy logic control today has attracted the interest of many researchers due to its cheap development, wide coverage of operational conditions and simplicity in setting, and for these reasons, it has been increasingly used in the control of HIV/AIDS recently [32–35]. In [32] a fuzzy-logic-based controller is proposed for the treatment of HIV-1-infected patients in which the imperialist competitive algorithm is designed for optimizing the designed coefficients of the controller. Fuzzy logic has been used in [33] to analyze and examine the structure of complex biological and epidemic models with an emphasis on HIV/AIDS. In [34], two different strategies named reverse transcriptase inhibitors (RTI) and protease inhibitors (PI) are used to treat HIV infection, in which a fuzzy aggregator of optimal type 2 also modifies the output of the controllers according to some optimal rules.

In this paper, after presenting a new nonlinear model for HIV/AIDS system, we use a fuzzy logic, and Adaptive method for solving the challenging characteristic of HIV medicines, and evaluation of the control techniques and any other new methods. In this paper, a new combined controller is made to catch the treatment up with the adaptive method. However, to check and analyze the proposed nonlinear models of this viral disease some clinical data have been also used properly. The main contributions of this paper can be highlighted as follows:

1. *Providing a new model for HIV/AIDS system based on healthy cells and infected cells.*
2. *Designing an adaptive control law with the method of self-regulating regulators compatible with the nature of the presented HIV/AIDS model.*
3. *Designing a Hybrid Adaptive-Fuzzy control in which the necessary dose of the drug is issued to the infected viruses to control the virus production rate in the blood.*

The article structure is illustrated in turn. In the second part, model dynamics of the system would be discussed. Then it gives the computation needed to design an adaptive controller. So, the robust fuzzy-adaptive controller is discussed. And the last but not least, conclusions and suggestions would be listed.

2 Dynamic Modeling of HIV/AIDS Infection

The model applied in this paper is a third-order model which includes intact impatient lymphocytes T, defective lymphocytes T and the other free viruses as state space variables. A general form of the system dynamics is revealed below.

Table 1 The dynamic parameters of HIV/AIDS model in (1)

Parameters	Description	Unit
s	CD4 + T, healthy cells regenerate rate	cell.mm^{-3}.day^{-1}
d	CD4 + T, healthy cell death rate	cell.mm^{-3}.day^{-1}
β	Infection rate coefficient	particle^{-1}.mm^{3}.day^{-1}
μ	CD4 + T, average death rate of healthy cells	day^{-1}
K	Coefficient of virus production	particle.cell$-^{1}$.day^{-1}
C	Average virus death rate	day^{-1}
U	Effect of PTI and RTI	–

$$\begin{cases} \dot{T} = s - dT - \beta T V \\ \dot{T}^{*} = \beta T V - \mu T^{*} \\ \dot{V} = (1-u)KT^{*} - cV \end{cases} \tag{1}$$

where T is the number of intact healthy lymphocytes T, S is a marker of the rate production of current free lymphocytes T in the bloodstream, V is free viruses in the blood and d is the rate of death of normal cells. Also, healthy cells at a specified rate β might be infected by the free viruses. In the second equation T^{*} is a marker of infected cells lymphocyte T that by a fixed rate specified μ could be destroyed gradually. In the third one, free viruses proliferate with the rate of K from infected cells and then eliminate at the fixed rate of c. In the third equation, the effect of drugs combining is indicated by u. If patients take medication successfully, infected cells will no longer be able to proliferate. Available parameters in the above equations with their measurement unit are mentioned in Table 1.

The variables T, T^{*} and V are the state variables of the mentioned system in (1) which are considered as a state vector as follows:

$$X = (x_1, x_2, x_3)' = (T, T^{*}, V)' \tag{2}$$

and the existing parameters of the model are defined as follows

$$\theta = (s, d, \mu, \beta, k, c) \tag{3}$$

Therefore, it can be revealed that

$$\dot{X} = f(X, \theta, u) \tag{4}$$

Before virus infection in the body, we have $x_2 = x_3 = 0$ and therefore, the number of impatient cells that desire to be at equilibrium point of $x_1 = \frac{s}{d}$, by virus entrance to the body the number of viruses goes to v_0. Therefore, the initial conditions for a contaminated person are given by

$$\begin{cases} x_1(0) = \dfrac{s}{d} \\ x_2(0) = 0 \\ x_3(0) = v_0 \end{cases} \tag{5}$$

For the infection's duration, measuring the original fertility ratio is important, R_0 considers as some new infected cells while almost all of the other cells are intact. So, it can be calculated as

$$R_0 = \frac{\beta k}{\mu c} x_1(0) = \frac{\beta k s}{\mu c d} \tag{6}$$

If $R_0 < 1$ then the infection would not be spread and on average, each infected cell produces less than another one. On the other hand, if $R_0 > 1$ the infection will be succeeded in proliferation and viruses are able to increase exponentially. Another important issue in the system model analysis is the equilibrium point of model. The model mentioned for the immunodeficiency virus includes two equilibrium points as shown below:

$$\begin{cases} \overline{X}^1 = \left(\dfrac{s}{d}, 0, 0\right) \\ \overline{X}^2 = \left(\dfrac{\mu c}{\beta k}, \dfrac{s}{\mu} - \dfrac{dc}{\beta k}, \dfrac{sk}{\mu c} - \dfrac{d}{\beta}\right) \end{cases} \tag{7}$$

Sample amounts of the parameters listed are as follows:

$$\theta = (s, d, \mu, \beta, k, c)' = (9, 0.009, 4 \times 10^{-6}, 0.3, 80, 0.6)' \tag{8}$$

where the initial conditions of system is

$$X_0 = X(0) = [1000, 0, 1]^T \tag{9}$$

At first, assume no medications have been taken, therefore for $u = 0$, the system state variable behaviors were analyzed. Figure 1 clearly shows the behavior of state variables of HIV model in (1) in the absence of any controller, under above mentioned initial conditions.

The results revealed virus exponentially proliferating in the early stage of infections accompanied with rapid decline of intact cells of lymphocyte T. After the first maximum pick, the number of intact cells of lymphocyte T and virus load led to the second equilibrium point which provided predicting for possible treatment probably. So,

$$X_{eq} = [562.5, 13.125, 1750]^T \tag{10}$$

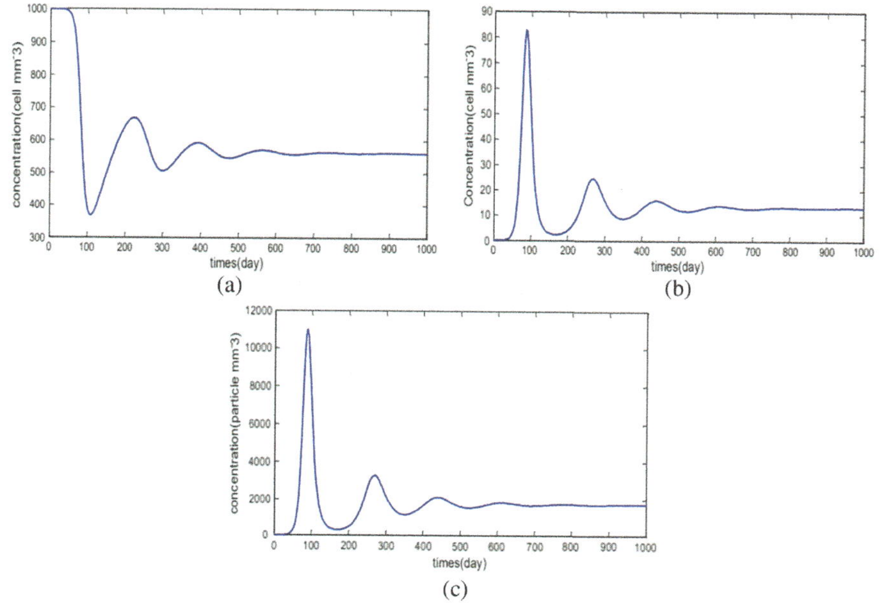

Fig. 1 State variable of HIV model in in open loop condition. **a** Uninfected cells, **b** Infected cells **c**. Free viruses in the blood

3 Adaptive Control Design to the Immunodeficiency Virus Model

In this section, using one of the adaptive control methods called self-tuning regulators, the design of the controller for the HIV/AIDS model is discussed.

Self-Tuning Regulators (STR) are created by combining the Recursive Least-Squares (RLS) estimator with the Minimum Degree Pole Placement (MDPP) method. In this control method, every time it is repeated, the identifier algorithm identifies the process parameters. Also, Diophantine Equations are solved based on the parameters estimated up to that moment. According to this procedure, at each stage when the parameters converge to their actual values, the controller parameters are also updated and control the process by generating a control signal based on the control law. Firstly, immunodeficiency virus model was linearized:

$$f(x,\theta) = \begin{cases} \dot{T} = s - dT - \beta TV \\ \dot{T}^* = \beta TV - \mu T^* \\ \dot{V} = (1-u)KT^* - cV \end{cases} \tag{11}$$

Considering the equilibrium point as in (10), we have

$$\begin{cases} X_1 = T - 562.5 \Rightarrow T = X_1 + 562.5 \\ X_2 = T^* - 13.125 \Rightarrow T^* = X_2 + 13.125 \\ X_3 = V - 1750 \Rightarrow V = X_3 + 1750 \end{cases} \quad (12)$$

Replacing the above relations to the HIV/AIDS system model in (1) one can obtain

$$\begin{cases} \dot{T} = s - d(X_1 + 562.5) - \beta(X_1 + 562.5)(X_3 + 1750) \\ \dot{T}^* = \beta(X_1 + 562.5)(X_3 + 1750) - \mu(X_2 + 13.125) \\ \dot{V} = (1 - u)k(X_2 + 13.125) - c(X_3 + 1750) \end{cases} \quad (13)$$

So the linearized system state space equation is equal to:

$$A = \frac{\partial f}{\partial X} = \begin{bmatrix} -d - \beta V & 0 & -\beta T \\ \beta V & -\mu & \beta T \\ 0 & k & -c \end{bmatrix}, B = \frac{\partial f}{\partial u} = \begin{bmatrix} 0 \\ 0 \\ -13.125k \end{bmatrix},$$

$$C = \begin{bmatrix} 1\,0\,0 \end{bmatrix}, D = \begin{bmatrix} 0 \\ 0 \\ 0 \end{bmatrix} \quad (14)$$

And its transfer function $G(s) = C(SI - A)^{-1}B + D$ is also given by

$$G(s) = \frac{TT^*\beta kS + TT^*\beta k\mu}{S^3 + (c + d + \mu + \beta V)S^2 + (cd + c\mu + d\mu + V\beta c - T\beta k + V\beta\mu)S + (cd\mu - Td\beta k + V\beta c\mu)} \quad (15)$$

Substituting the parameter's value inside the transfer function in (15) results

$$G(S) = \frac{236250S + 70875}{10^5 S^3 + 91600S^2 + 1440S + 126} \quad (16)$$

Discrete form of the mentioned transfer function with $t_s = 2$ as time sampling is

$$G(q) = \frac{3.412q^2 + 0.451q - 1.276}{q^3 - 2.131q^2 + 1.296q - 0.1601} \quad (17)$$

The third-order system could be estimated to second one as follows:

$$G(q) = \frac{B(q)}{A(q)} = \frac{b_0 q + b1}{q^2 + a_1 q + a_2}, G(q) = \frac{y(t)}{u(t)} = \frac{b_0 q^{-1} + b_1 q^{-2}}{1 + a_1 q^{-1} + a_2 q^{-2}} \quad (18)$$

Using the procedure of designing the STR adaptive control, at first, the parameters of system above would be achieved by Regressive Least Square (RLS) method as follows:

$$y(t) = -a_1 y(t-1) - a_2 y(t-2) + b_0 u(t-1) + b_1 u(t-2) \tag{19}$$

In the next turn, an acceptable reference model might be chosen. As considering the order of system, the order of reference model could be two indeed. This model is a desirable model according to the situation of HIV in a specific region.

$$G_m(q) = \frac{b_m(q)}{A_m(q)} = \frac{b_{m0}q + b_{m1}}{q^2 + a_{m1}q + a_{m2}} \tag{20}$$

By assuming the amounts of $\xi = 0.9$, $w_n = 0.04$ for second order reference model, it could be illustrated as

$$\begin{aligned} G_m(s) &= \frac{\omega_n^2}{s^2 + 2\xi\omega_n s + \omega_n^2} = \frac{0.0016}{s^2 + 0.0185s + 0.0016}, \\ G_m(q) &= \frac{0.0305q + 0.002907}{q^2 - 1.86q + 0.8659} \end{aligned} \tag{21}$$

Typical form of the transfer function of control law is given in below

$$R(q)u(t) = T(q)u_c(t) - S(q)y(t),\ S(q) = s_0 q + s_1,\ R(q) = q + r_1 \tag{22}$$

According to causal property of the system, reference model and controller, the parameters degree could be estimated as follows:

$$\begin{aligned} &\deg(A) + \deg(R) = \deg(A_m) + \deg(A_o) \text{ And } \deg(A) = 2, \deg(R) = 1, \\ &\deg(A_m) = 2, \deg(A_0) = 1 \text{ And } \Rightarrow A_o = q + a_{o1} \end{aligned} \tag{23}$$

For finding the amounts of controller parameters, Diophantine Equations meaning $AR + BS = A_m A_o$ should be solved as below. By considering the reached conditions in the above equations and solving the Diophantine Equation, one can obtain

$$(q^2 + a_1 q + a_2)(q + r_1) + (b_0 q + b_1)(s_0 q + s_1) = (q^2 + a_{m1}q + a_{m2})(q + A_{o1}) \tag{24}$$

which results in

$$\begin{cases} r_1 + a_1 + b_0 s_0 = a_{o1} + a_{m1} \\ a_1 r_1 + a_2 + b_1 s_0 + b_0 s_1 = a_{o1} a_{m1} + a_{m2} \\ a_2 r_1 + b_1 s_1 = a_{o1} a_{m2} \end{cases} \approx \begin{bmatrix} 1 & b_0 & 0 \\ a_1 & b_1 & b_0 \\ a_2 & 0 & b_1 \end{bmatrix} \begin{bmatrix} r_1 \\ s_0 \\ s_1 \end{bmatrix} = \begin{bmatrix} a_{o1} + a_{m1} - a_1 \\ a_{o1} a_{m1} + a_{m2} - a_2 \\ a_{o1} a_{m2} \end{bmatrix} \quad (25)$$

Also, for finding T, we should use the equations $T = \beta A_o$, $\beta = A_m(1)/B(1)$ and

$$T(q) = \left(\frac{1 + a_{m1} + a_{m2}}{b_0 + b_1}\right)(q + a_{o1}) = \left(\frac{1 + a_{m1} + a_{m2}}{b_0 + b_1}\right)q + \left(\frac{a_{o1} + a_{m1} a_{o1} + a_{m2} a_{o1}}{b_0 + b_1}\right), \quad (26)$$

That clearly results in

$$t_0 = \frac{1 + a_{m1} + a_{m2}}{b_0 + b_1}, t_1 = \frac{a_{o1} + a_{m1} a_{o1} + a_{m2} a_{o1}}{b_0 + b_1} \quad (27)$$

For designing real controller, the relation between R, S and T could be written with respect to shift delay q factor as $R(q) = q + r_1 \Rightarrow R(q)q^{-1} = 1 + r_1 q^{-1}$, $S(q) = s_0 q + s_1 \Rightarrow S(q)q^{-1} = s_0 + s_1 q^{-1}$ and $T(q) = t_0 q + t_1 \Rightarrow T(q)q^{-1} = t_0 + t_1 q^{-1}$. Finally, by replacing the above equations in the main controller $R(q)u(t) = T(q)u_c(t) - S(q)y(t)$, it yields:

$$(1 + r_1 q^{-1})u(t) = (t_0 + t_1 q^{-1})u_c(t) - (s_0 + s_1 q^{-1})y(t) \quad (28)$$

which can be written as

$$u(t) = -r_1 u(t-1) + t_0 u_c(t) + t_1 u_c(t-1) - s_0 y(t) - s_1 y(t-1) \quad (29)$$

Now, we are ready to show the results of applying the designed adaptive control for stabilizing the HIV/AIDS model in (1). The simulation results are shown in Figs. 2 and 3.

By adjusting the parameters of the least squares algorithm and setting the polynomial observer, the desired answers per square input are observed. We know the system input is between zero and one, so the configurable parameters are changed so that u does not go out of this range (Fig. 2).

Now it is time for the main step, which is to place the nonlinear model instead of the linear model. After setting the parameters, the desired response was observed. Note that the nonlinear system near the equilibrium point acts like a linear system as in Fig. 3.

In these articles, the goal is to reach 1000 healthy cells, with this in mind, adjustment has been made. The figure above shows that with this design, the number of

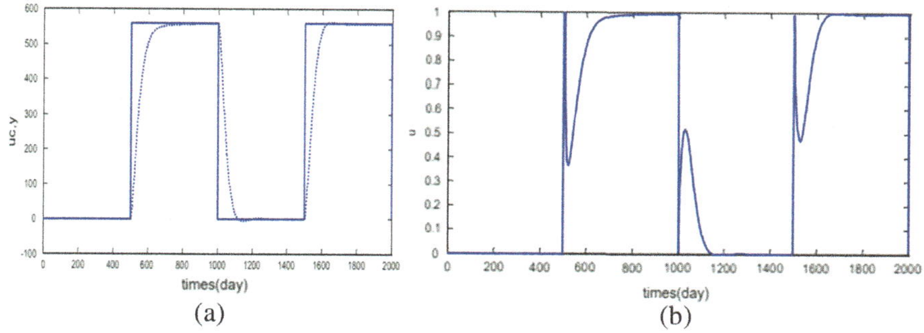

Fig. 2 Simulation results of using Adaptive control for HIV/AIDS linear model. **a** Draw the input Uc and output Y according to the number of days of the linear system, **b** U input for linear system (drug effect)

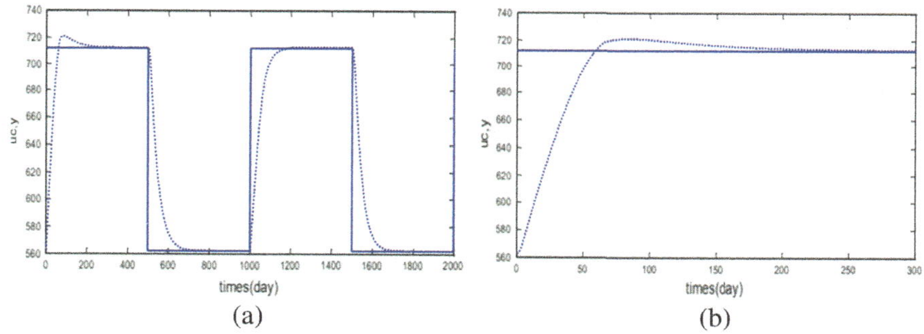

Fig. 3 Simulation results for nonlinear model of HIV/AIDS system in (1). **a** Input uc and output y in terms of number of days for real nonlinear system, **b** Convergence in first days

healthy cells has finally reached 732. In Fig. 3, the input is between zero and one and is not out of this range (Fig 3a).

As shown in Fig 3b, the system is slow and after about 611 days' increases from 562 cells to 732 cells, which is not enough to cure the patient. Therefore, a method should be expressed so that the number of healthy cells in the patient's body reaches 1000 per cubic millimeter, in order to provide conditions for recovery. In the next part, a combination of adaptive controller and fuzzy control will solve this problem.

4 Using Adaptive-Fuzzy Control Method

Use of fuzzy logic returns to Lotfizade's theorem that is based on experimental knowledge. Therefore, it's very useful even for those complicated systems with bad structures that designers can implement this suitable experimental method very well, and even for other simple structures. For example, if you can't linearize a system using a linear controller, fuzzy logic helps you. In this part application of a Fuzzy Logic Controller to the highly nonlinear and unstable HIV system is presented. At first, an initial input is applied to the viruses to trig the disease and make the system unbalanced, which makes the disease error with respect to the desired position, then it is the controller's responsibility to catch the treatment up and stabilized it. Therefore, the controller should decrease the disease error to zero. The fuzzy controller should use the real state variables, the HIV outputs are evaluated then we should select the suitable behavior of those outputs and at the end with using the controllability parameters to control the unwanted behavior of the disease to lead to our desirable response which is to keep viruses near the zero. Our FLC inputs are amounts of infected and unhealthy cells with its initial conditions that go to the rule base using the fuzzy rules that are experimental rules refer to our experience about the real system. Using these commands, a necessary dosage of medicines is exported to the viruses to control the rate of generation of viruses in blood. The FLC output which is the dosage of medicines goes to the u signal to implement fuzzy commands. In this case, the commanded dosage uses the T cells and anti-viral as we want to return the falling cells to a healthy alive position or contrariwise for balancing it to zero rates. In the following picture, the fuzzy rule base is shown. Therefore, in this paper it is shown that we can use a fuzzy controller instead of the usual single controllers, we could have a better performance of the implementation of fuzzy logic using our proposed method accompanied with an adaptive one. Fuzzy logic may not be a very fast controller to have a quick response because of its highly nonlinear behavior, but it's very useful for most systems and also for complicated systems because of design difficulties (Fig. 4).

In order to use Fuzzy-Adaptive Control for HIV model and using $X = [T, T^*, V]^T$ as state variables, Differential equation of immunodeficiency virus model could be given as follows in input affine format:

$$x^{(n)} = f(x) + g(x)u \tag{30}$$

where

Fig. 4 Fuzzy surface of rules memberships

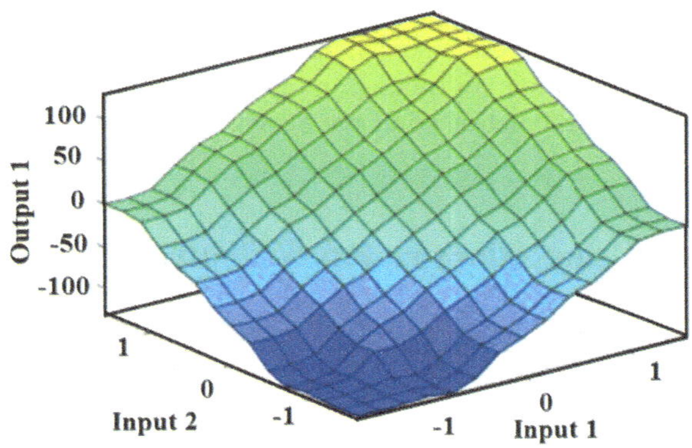

$$f(x) = \begin{bmatrix} s - dT - \beta TV \\ \beta TV - \mu T^* \\ kT^* - cV \end{bmatrix}, \text{ and } g(x) = \begin{bmatrix} 0 \\ 0 \\ -kT^* \end{bmatrix}$$

This nonlinear considered system have $f(x)$ and $g(x)$ as nonlinear terms that include vector function that can be linearized such as a single input system smoothly.

Linearizing of state input- could be shown below:

$$ad_{f^0}g = \begin{bmatrix} 0 \\ 0 \\ -kT^* \end{bmatrix} = g \tag{31}$$

$$ad_f g = \begin{bmatrix} \beta cTV - \beta kTT^* \\ -\beta cTV + \beta kTT^* \\ k\mu T^* - k\beta TV \end{bmatrix} = F_1 \tag{32}$$

$$ad_{f^2}g = \begin{bmatrix} \beta\big(csV - KsT^* - c^2TV - \beta cT^2V + \beta kT^2T^* \\ -\beta kT^2V + ckTT^* + k\mu TT^*\big) \\ -\beta(csV - ksT^* - c^2TV - \beta cT^2V + \beta kT^2V \\ -cdTV + ckTT^* + dkTT^* + c\mu TV \\ \beta\big(cV - kT^*\big)(dT - s + \beta TV) + \beta kT\big(cV - kT^*\big) \\ -\beta kT\big(T^*\mu - \beta TV\big) \end{bmatrix} = F_2 \tag{33}$$

Controllability Matrix is given below:

$$\big[g \; ad_f g \; ad_{f^2}g\big] = \big[g \; F_1 \; F_2\big]$$

where the following assumptions are considered

$$L_g z_1 = L_{ad_f g} z_1 = 0, \; L_{ad_{f^2}g} z_1 \neq 0, \; \frac{\partial z_1}{\partial T} = \frac{\partial z_1}{\partial T^*}, \frac{\partial z_1}{\partial V} = 0 \tag{34}$$

So the new variables z_1, z_2 are evaluated like below:

$$z_1 = T + T^* - T_{eq} - T^*_{eq} \tag{35}$$

$$z_2 = \nabla z_1 f = (s - dT - \beta TV) + (BTV - \mu T^*) = s - dT - \mu T^* \tag{36}$$

So z_2, z_3 are made too

$$\begin{aligned} z_3 &= \nabla z_2 f = -d(s - dT - \beta TV) - \mu(\beta TV - \mu T^*) \\ &= -sd + d^2T + d\beta TV - \mu\beta TV + \mu^2 T^* \Rightarrow z_3 \\ &= -sd + d^2T + d\beta TV - \mu\beta TV + \mu^2 T^* \end{aligned} \tag{37}$$

According to parameters values and initial conditions, the equilibrium point of system is now computed as follows

$$\begin{aligned} z_{1eq} &= T_{eq} + T^*_{eq} - T_{eq} - T^*_{eq} = 0 \\ z_{2eq} &= s - dT_{eq} - \mu T^*_{eq} = 0 \\ z_{3eq} &= -sd + d^2T_{eq} + d\beta T_{eq}V_{eq} - \mu\beta T_{eq}V_{eq} + \mu^2 T^*_{eq} = 0 \end{aligned} \tag{38}$$

As a result, the essential existence of equilibrium point $z_{eq} = [0, 0, 0]^T$ is proved rightly. In the next step variable changes of states such as T, T^*, V, z_1, z_2, z_3 would be analyzed in turn. Using the bellowing equations

$$\begin{cases} z_1 = T + T^* - T_{eq} - T^*_{eq} \\ z_2 = s - dT - \mu T^* \end{cases} \Rightarrow \begin{cases} \mu z_1 = \mu T + \mu T^* - \mu T_{eq} - \mu T^*_{eq} \\ z_2 = s - dT - \mu T^* \end{cases}$$

Then the variable T can be obtained as

$$T = \frac{z_2 + \mu z_1 - s + \mu(T_{eq} + T^*_{eq})}{(\mu - d)} \tag{39}$$

And using (35) and (39), one can obtain

$$\begin{aligned} T^* &= z_1 - \frac{z_2 + \mu z_1 - s + \mu(T_{eq} + T^*_{eq})}{(\mu - d)} + (T_{eq} + T^*_{eq}) \Rightarrow T^* \\ &= z_1 - T + T_{eq} + T^*_{eq} \end{aligned} \tag{40}$$

Which is simplified as

$$T^* = \frac{-z_1 d - z_2 - \mu z_1 + s - d(T_{eq} + T^*_{eq})}{(\mu - d)} \tag{41}$$

Also, it is achieved that:

$$\begin{aligned} z_3 &= -sd + d^2T + d\beta TV + \mu\beta TV + \mu^2 T^* \\ &= -sd + d^2\left(\frac{z_2 + \mu z_1 - s + \mu(T_{eq} + T^*_{eq})}{(\mu - d)}\right) \\ &\quad + \beta\left(\frac{z_2 + \mu z_1 - s + \mu(T_{eq} + T^*_{eq})}{(\mu - d)}\right)V \\ &\quad + \mu^2\left(\frac{-z_1 d - z_2 - \mu z_1 + s - d(T_{eq} + T^*_{eq})}{(\mu - d)}\right) \end{aligned} \tag{42}$$

where results

$$V = \frac{z_3(\mu - d) + sd(\mu - d)}{\beta(z_2 + \mu z_1 - s + \mu(T_{eq} + T^*_{eq}))} + \frac{-d^2(z_2 + \mu z_1 - s + \mu(T_{eq} + T^*_{eq})) + \mu^2(-z_1 d - z_2 - \mu z_1 + s - d(T_{eq} + T^*_{eq}))}{\beta(z_2 + \mu z_1 - s + \mu(T_{eq} + T^*_{eq}))} \tag{43}$$

Now, derivative of (35) with respect to time and considering that the derivate of equilibrium points is zero, one can conclude

$$\begin{aligned} \dot{z}_1 &= \dot{T} + \dot{T}^* = (s - dT - \beta TV) + (\beta TV - \mu T^*) = s - dT - \mu T^* \\ \dot{z}_2 &= -d\dot{T} - \mu\dot{T}^* \Rightarrow \dot{z}_2 = -d(s - dT - \beta TV) - \mu(\beta TV - \mu T^*) \\ \dot{z}_3 &= d^2\dot{T} + d\beta\dot{T}V + d\beta T\dot{V} - \mu\beta T\dot{V} + \mu^2\dot{T}^* \end{aligned} \tag{44}$$

Now, considering a chain form as $\dot{z}_1 = z_2$ and $\dot{z}_2 = z_3$ and replacing the variables $\dot{T}$, $\dot{T}^*$, V and using some calculations yields:

$$\begin{aligned} \dot{z}_3 = \big(&sd^2 - d^3T - 2d^2\beta TV + ds\beta V - d\beta^2TV^2 + d\beta kTT^* - d\beta cTV \\ &- \mu s\beta V + \mu\beta dVT + \mu\beta^2TV^2 - \mu\beta kTT^* + \mu\beta cTV + \mu^2\beta TV - \mu^3T^*\big) \\ &+\big(-d\beta kTT^* + \mu\beta kTT^*\big)u \end{aligned} \tag{45}$$

The above relations along with equations in (44), can be rewritten as follows

$$\begin{aligned} \dot{z}_1 &= z_2 \\ \dot{z}_2 &= z_3 \\ \dot{z}_3 &= f(Z) + g(Z)u + d(t) \end{aligned} \tag{46}$$

According to achieved equations in (46) we are now ready to design a new Fuzzy-Adaptive Robust Controller by using new variables appropriately.

Model in (46) can be rewritten as below for the first variable

$$\begin{aligned} \dot{z}_1 &= z_2 \\ \ddot{z}_1 &= z_3 \\ \dddot{z}_1 &= f(Z) + g(Z)u + d(t) \end{aligned} \tag{47}$$

So, the companion model would be changed to: $\dddot{z} = f(Z) + g(Z)u + d(t)$. Here the notation $d(t)$ is considered to be disturbance in HIV model.

Now, we are ready to propose the following control law for stabilizing the HIV model in (47)

$$\Rightarrow u = \frac{\dddot{z}_d + k_1(\ddot{z}_d - \ddot{z}) + k_2(\dot{z}_d - \dot{z}) + k_3(z_d - z) - \hat{f}(Z) - u_r}{\hat{g}(Z)} \tag{48}$$

where $u_r = \hat{W}\mathrm{sgn}(E^T SB)$.

By subtracting (48) from (47), one can obtain

$$\dddot{e} + k_1\ddot{e} + k_2\dot{e} + k_3 e = (\hat{f}(Z) - f(Z)) + (\hat{g}(Z) - g(Z))u + d(t) - u_r \tag{49}$$

In the event that the design is done correctly, then

$$\begin{aligned} &\dddot{e} + k_1\ddot{e} + k_2\dot{e} + k_3 e = 0 \\ &e \to 0 \Rightarrow z_d - z \to 0 \Rightarrow z_d \to z \end{aligned} \tag{50}$$

The controller system functions $f(Z)$ and $g(Z)$ could be estimated in fuzzy format as follows

$$\begin{aligned} \hat{f}(Z) &= \hat{p}^T \xi_f(Z) \\ \hat{g}(Z) &= \hat{q}^T \xi_g(Z) \end{aligned} \tag{51}$$

which results in the following error equation

$$\dddot{e} + k_1\ddot{e} + k_2\dot{e} + k_3 e = (\hat{p}^T - p^T)\xi_f(Z) + (\hat{q}^T - q^T)\xi_g(Z)u + d(t) - u_r$$

where

$$\begin{aligned} p^* &= \arg\min_{p\in\Omega_p} \sup\nolimits_{x\in R^n} \left\| f(X) - \hat{f}(x|p) \right\| \\ q^* &= \arg\min_{q\in\Omega_q} \sup\nolimits_{x\in R^n} \left\| g(X) - \hat{g}(x|q) \right\| \end{aligned} \tag{52}$$

By considering the following definitions, the minimum approximation error (W) could be obtained using following equations

$$\begin{cases} \tilde{p} = \hat{p} - p^* \\ \tilde{q} = \hat{q} - q^* \\ W = (p^{*T} - p^T)\xi_f(Z) + (q^{*T} - q^T)\xi_g(Z)u + d(t) \end{cases} \tag{53}$$

Which results in the following new error equation as closed loop system

$$\dddot{e} + k_1\ddot{e} + k_2\dot{e} + k_3 e = \tilde{p}^T \xi_f(Z) + \tilde{q}^T \xi_g(Z)u + W - u_r \tag{54}$$

$$\Rightarrow \begin{cases} \dot{e}_1 = e_2 \\ \dot{e}_2 = e_3 \\ \dot{e}_3 = -k_1 e_3 - k_2 e_2 - k_3 e_1 + \tilde{p}^T \xi_g(Z)u + W - u_r \end{cases}$$

The matrix format of equations in (54) is like $\dot{E} = AE + B(\tilde{p}^T \xi_f(Z) + \tilde{q}_g(Z)u + w - u_r)$ where

$$A = \begin{bmatrix} 0 & 1 & 0 \\ 0 & 0 & 1 \\ -k_3 & -k_2 & -k_1 \end{bmatrix}, B = \begin{bmatrix} 0 \\ 0 \\ 1 \end{bmatrix} \tag{55}$$

Now, we are ready to prove the stability of the closed loop system using Lyapunov theory. To do this, first consider the following Lyapunov candidate function

$$V = \frac{1}{2}E^T SE + \frac{1}{2\alpha}(\hat{p}^T - p^{*T})(\hat{p} - p^*) + \frac{1}{2\beta}(\hat{q}^T - q^{*T})(\hat{q} - q^*) + \frac{1}{2\gamma}\hat{W}^2 \tag{56}$$

where $\tilde{W} = \hat{W} - W_m$ such that $|W| \leq W_m$. W_m is considered as upper bound of term W and it is clear that W is unknown because of the existence of unknown disturbance $d(t)$ in it as in (53).

Now, differentiating of Lyapunov function in (56) results

$$\begin{aligned} \dot{V} &= \frac{1}{2}\dot{E}^T SE + \frac{1}{2}E^T S\dot{E} + \frac{1}{2\alpha}(\dot{\hat{p}}^T)(\hat{p} - p^*) + \frac{1}{2\alpha}(\hat{p}^T - p^{*T})(\dot{\hat{p}}) \\ &+ \frac{1}{2\alpha}(\dot{\hat{q}})(\hat{q} - q^*) + \frac{1}{2\beta}(\hat{q}^T - q^{*T})(\dot{\hat{q}}) + \frac{2}{2\gamma}\tilde{W}\dot{\hat{W}} \end{aligned} \tag{57}$$

Assume $A^T S + SA = -Q$ where Q is an arbitrary symmetrical positive definition and then from (57) we have

$$\begin{aligned} \dot{V} &= -\frac{1}{2}E^T QE + E^T SBW + \left(E^T SB\xi_f(Z) + \frac{1}{\alpha}\dot{\hat{p}}\right)(\dot{\hat{p}} - p^{*T}) \\ &+ \left(E^T SB\xi_g(Z)u + \frac{1}{\beta}\dot{\hat{q}}\right)(\hat{q}^T - q^{*T}) - E^T SBu_r + \frac{1}{\gamma}\tilde{W}\dot{\hat{W}} \end{aligned} \tag{58}$$

The upper bound of the derivative of Lyapunov in (58) can be rewritten as follows

$$\begin{aligned} \dot{V} &\leq -\frac{1}{2}E^T QE + \left(E^T SB\xi_f(Z) + \frac{1}{\alpha}\dot{\hat{p}}\right)(\hat{p}^T - p^{*T}) \\ &+ \left(E^T SB\xi_g(Z)u + \frac{1}{\beta}\dot{\hat{q}}\right)(\hat{q}^T - q^{*T}) + |E^T SB|W_m - |E^T SB|\hat{W} \\ &+ \frac{1}{\gamma}\tilde{W}\dot{\hat{W}} \leq -\frac{1}{2}E^T QE + \left(E^T SB\xi_f(Z) + \frac{1}{\alpha}\dot{\hat{p}}\right)(\hat{p}^T - p^{*T}) \\ &+ \left(E^T SB\xi_g(Z)u + \frac{1}{\beta}\dot{\hat{q}}\right)(\hat{q}^T - q^{*T}) - |E^T SB|\tilde{W} + \frac{1}{\gamma}\tilde{W}\dot{\hat{W}} \end{aligned} \tag{59}$$

Now, consider the following equations in order to cancel some terms in right hand side of (59)

$$\begin{aligned} &E^T SB\xi_f(Z) + \frac{1}{\alpha}\dot{\hat{p}} = 0 \\ &E^T SB\xi_g(Z) + \frac{1}{\beta}\dot{\hat{q}} = 0 \\ &\frac{1}{\gamma}\dot{\hat{W}} - \left|E^T SB\right| = 0 \end{aligned} \tag{60}$$

where equations in (60) are actually the adaptive laws for obtaining controller parameters and also the upper bound of the disturbance term as follows

$$\begin{aligned} &\dot{\hat{p}} = -\alpha E^T SB\xi_f(Z) \\ &\dot{\hat{q}} = -2\alpha E^T SB\xi_g(Z)u \\ &\dot{\hat{W}} = \gamma\left|E^T SB\right| \end{aligned} \tag{61}$$

Using (61), the inequality in (59) can be written as follows

$$\dot{V} \le -\frac{1}{2}E^T QE \tag{62}$$

So $\dot{V}$ is negative semi defines, and given that the derivation of $\dot{V}$ could be shown that $\ddot{V}$ is also bounded, then according to Barbalat Lemma it is revealed that: $\lim_{t\to\infty} \dot{V} = 0$ which is equal to $\lim_{t\to\infty} E = 0$ and this completes the proof. ■

Using (53) and (61), it is also concluded that $\xi(Z)$ is as follows

$$\xi(Z) = \frac{\prod_{i=1}^{n} \mu_{A_i^{li}}(Z_i)}{\sum_{l=1}^{m}\prod_{i=1}^{n} \mu_{A_i^{li}}(Z_i)} \tag{63}$$

In order to design an adaptive fuzzy controller, it is done in such a way that first the main system of HIV/AIDS is combined with the obtained transfer functions, then the control rule is designed, and finally it is applied to the main system with the help of the inverse transfer function, which presents the following results.

As a result, with this comparison, we find that this method has been very successful in the treatment process of HIV/AIDS disease. Figure 5 shows the effect of the drug in two methods. First, fuzzy-adaptive control method is presented in this paper and second the optimal controller is designed in reference [23]. As it is clear, the amplitude of input controller that stands for the drug in the fuzzy-adaptive controller is significantly less which shows the superiority of the fuzzy-adaptive control method for stabilizing the HIV/AIDS model.

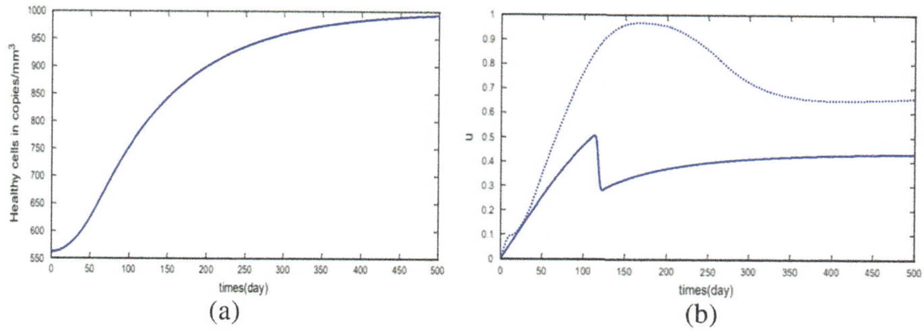

Fig. 5 Simulation results for nonlinear model of HIV/AIDS system using fuzzy-adaptive controller in (1). **a** Population of healthy cells in the blood, **b** Comparison of the effect of drug in adaptive fuzzy method with comparative to reference in [23]

5 Conclusion

In this paper, HIV/AIDS disease modeling and control is discussed. In this sense, first, a mathematical model based on the number of healthy and infected cells and ready to use drug is presented. Then, an adaptive control law with the title of self-tuning regulators is designed and presented for stabilizing the HIV/AIDS model. Then, first, the system model is converted into a special chain form by considering external disturbances and a new fuzzy-adaptive control law has been designed to stabilize the HIV/AIDS system, taking into account the minimum approximate error, and using Lyapunov to prove the stability. In this paper, adaptive rules for estimation of fuzzy controller parameters as well as upper bound estimation of disturbance terms are presented. The simulation results of each section show the correctness of the presented model and designed controllers.

References

1. UNAIDS (2010) UNAIDS report on the global AIDS epidemic, UNAIDS, Technical report
2. Conte G, Moog CH, Perdon AM (2007) Algebraic methods for nonlinear control systems, 2nd edn. Springer, London, UK
3. Mhawej M, Moog CH, Biafore F, Brunet-Francois C (2010) Control of the HIV infection and drug dosage. Biomed Sig Process Control 5:45–52
4. Chang H, Shim H, Seo J (2004) Control of immune response of HIV infection model by gradual reduction of drug dose. In: 43rd conference on decision and control Atlantis, Paradise Island, Bahamas
5. Mhawej MJ, Moog CH, Biafore F (2009) Drug dosage control of the HIV infection dynamics. In: Proceedings of JOBIM, Nantes
6. Legrand M, Comets E, Aymard G, Tubiana R, Katlama C, Diquet B (2003) An in vivo pharmacokinetic, pharmacodynamic model for antiretroviral combination. HIV Clin Trials 4(3):170–183
7. Ljung L (1999) System identification: theory for the user, 2nd edn. Prentice Hall PTR

8. Perelson A, Nelson P (1999) Mathematical analysis of HIV-1 dynamics in vivo. SIAM 41(1):3–44
9. Rivadeneira PS, Moog CH (2012) Impulsive control of single-input nonlinear systems with application to HIV dynamics. Math Comput 218:8462–8474
10. Mirzadeh R, Zare A (2011) Fuzzy logic controls designed based on genetic algorithms for a model of disease HIV. In: 3rd Iranian conference on electrical and electronics engineering, ICEEE
11. Andrade Costa B, Lemos JM (2011) Nonlinear feedback control of a HIV-1 infection model. In: 19th mediterranean conference on control and automation aquis corfu holiday palace, Corfu, Greece, pp 20–23
12. Zarei H, Vahidian Kamyad A, Effati S (2011) An adaptive memetic algorithm with dynamic population management for designing HIV multidrug therapies. Int J Comput Sci 133–140
13. Teklu SW, Rao KP (2022) HIV/AIDS-pneumonia codynamics model analysis with vaccination and treatment. Comput Math Methods Med 3105734:1–20
14. Attaullah MS (2020) Mathematical modeling and numerical simulation of HIV infection model. Results Appl Math 7:100118
15. Ayele TK, Doungmo Goufo EF, Mugisha S (2021) Mathematical modeling of HIV/AIDS with optimal control: a case study in Ethiopia 26:104263
16. Raza A, Ahmadian A, Rafiq M, et al (2020) Modeling the effect of delay strategy on transmission dynamics of HIV/AIDS disease. Adv Diff Equ 2020:663
17. Sun X, Nishiura H, Xiao Y (2020) Modeling methods for estimating HIV incidence: a mathematical review. Theor Biol Med Model 17:1
18. Guerrero–Sánchez Y, Umar M, Sabir Z, Guirao JLG, Zahoor Raja MA (2021) Solving a class of biological HIV infection model of latently infected cells using heuristic approach. Discrete Continuous Dyn Syst 14(10):3611–3628
19. Elaiw AM, AlShamrani NH (2021) Modeling and analysis of a within-host HIV/HTLV-I co-infection. Boletín de la Sociedad Matemática Mexicana 27:38
20. Levy B, Correia HE, Chirove F, et al (2021) Modeling the effect of HIV/AIDS stigma on HIV infection dynamics in Kenya. Bull Math Biol 83:55
21. Soorki MN, Tavazoei MS (2013) Fractional-order linear time invariant swarm systems: asymptotic swarm stability and time response analysis'. Cent Eur J Phys 11(6):845–854
22. Naderi Soorki M, Tavazoei MS (2018) Adaptive robust control of fractional-order systems in the presence of model uncertainties and external disturbances. IET Control Theory Appl 12(7):961–969
23. Naderi Soorki M, Tavazoei MS (2014) Adaptive consensus tracking for fractional-order linear time invariant swarm systems. ASME J Comput Nonlinear Dyn 9(3):031012
24. Soorki MN, Moghaddam TV, Emamifard A (2021) A new fast finite time fractional order adaptive sliding-mode control for a quadrotor. In: 2021 7th international conference on control, instrumentation and automation (ICCIA), pp 1–5
25. Naderi Soorki M, Tavazoei MS (2016) Constrained swarm stabilization of fractional order linear time invariant swarm systems. IEEE/CAA J Autom Sinica 3(3):320–331
26. Aslam M, Murtaza R, Abdeljawad T et al (2021) A fractional order HIV/AIDS epidemic model with Mittag-Leffler kernel. Adv Diff Equ 2021:107
27. Habibah U, Trisilowati T, Tania TR, Alfaruq LU (2020) Control strategy of HIV/AIDS model with different stages of infection of subpopulation. In: Journal of physics: conference series, volume 1872, 1st international conference on mathematics and its applications (ICoMathApp), 30 Sept 2020, Malang, Indonesia, pp 1–7
28. Najafi M, Basirzadeh H (2022) A new optimal control technique for solution of HIV infection model. Boletim da Sociedade Paranaense de Matemática 40:1–7
29. Rabiu M, Willie R, Parumasur N (2021) Optimal control strategies and sensitivity analysis of an HIV/AIDS-resistant model with behavior change. Acta Biotheor 69(4):543–589
30. Jo NH (2019) Robust drug treatment for HIV-1 infection model with completely unknown parameters. Int J Control Autom Syst 17:3113–3121

31. Bera MK, Kumar P, Biswas RK (2019) Robust control of HIV infection by antiretroviral therapy: a super-twisting sliding mode control approach. IET Syst Biol 13(3):120–128
32. Reisi NA, Hadipour Lakmesari S, Mahmoodabadi MJ, Hadipour S (2019) Optimum fuzzy control of human immunodeficiency virus type1 using an imperialist competitive algorithm. Inform Med Unlocked 16:100241
33. Lefevr N, Kanavos A, Gerogiannis VC, Iliadis L, Pintelas P (2021) Employing fuzzy logic to analyze the structure of complex biological and epidemic spreading models. Mathematics 9(9):977
34. Chen ShB, Rajaee F, Yousefpour A et al (2021) Antiretroviral therapy of HIV infection using a novel optimal type-2 fuzzy control strategy. Alex Eng J 60(1):1545–1555
35. Ahsan M, Saeed M, Mehmood A, Saeed MH, Asad J (2021) The study of HIV diagnosis using complex fuzzy hypersoft mapping and proposing appropriate treatment. IEEE Access 9:104405–104417

Communication Networks Characteristics Impact on Cyber-Physical Systems

Mehdi Zeinali and Reza Vatankhah Barenji

Abstract This study is exploring the impact of data transmission in cyber-physical systems, which rely on the exchange of data between various hardware and software components. The study aims to identify the most effective practices for designing and implementing data transmission in these systems and how to apply them to real-world applications. The study analyzes cyber-physical systems' data transmission requirements and various methods for improving data transmission. It provides valuable insights for designing and implementing cyber-physical systems, with practical applications in the field of Industry 4.0. The results inform future designers on how to improve the performance and reliability of cyber-physical systems by employing appropriate communication.

Keywords Cyber-physical system · Data transmission · Industry 4.0

1 Introduction

Cyber-physical systems, also known as remote mechatronic systems, are rapidly becoming an integral part of modern automation, robotics, and industrial control applications. These systems rely on the exchange of data between various hardware and software components to achieve their desired functions, such as monitoring and controlling remote processes, transmitting information, and providing feedback. Given the importance of data transmission in remote mechatronic systems, it is essential to understand the impact of this critical component on the performance and reliability of these systems [1–2].

Data transmission in cyber-physical systems is a complex and challenging task that is influenced by several factors, including transmission rate, distance, environmental conditions, and the complexity of the data being transmitted. For example, transmission rates that are too slow may result in a degradation of system performance, while high transmission rates may lead to increased energy consumption and reduced

M. Zeinali (✉) · R. V. Barenji
School of Science and Technology, Nottingham Trent Univesity, Engineering Department, Nottingham, UK
e-mail: Mehdi.Zeinali@ntu.ac.uk

© The Author(s), under exclusive license to Springer Nature Singapore Pte Ltd. 2023

A. Azizi (ed.), *Control Engineering in Mechatronics*, Emerging Trends in Mechatronics,
https://doi.org/10.1007/978-981-16-7775-5_6

battery life in mobile systems. Environmental conditions, such as temperature and humidity, may also affect data transmission, leading to errors and degradation in system performance [2–4].

In addition to these challenges, cyber-physical systems often operate in harsh and dynamic environments, where the availability of communication channels is limited and the reliability of data transmission is critical. The data transmitted by these systems often contain time-sensitive information that must be transmitted in real-time, such as sensor readings and control signals. In these cases, errors or disruptions in data transmission may result in system failure and even safety hazards. Therefore, it is important to understand the impact of data transmission on the performance and reliability of CPSs and to develop methods for improving data transmission.

This study aims to examine the impact of data transmission in cyber-physical systems, focusing on the following research question: what are the best practices for designing and implementing data transmission in cyber-physical systems, and how can these best practices be applied to real-world applications? To answer this research question, this study will analyze the data transmission requirements of cyber-physical systems and the various methods for improving data transmission. The study will provide valuable insights into the design and implementation of CPSs and will have practical applications in the fields of Industry 4.0. The results of this research will inform the future designer that can improve the performance and reliability of their CPS.

The structure of this chapter is as follows: In the next section, we will provide a background on cyber-physical systems and data transmission. This will be followed by a review of the relevant literature on the impact of data transmission in cyber-physical systems. In the subsequent sections, we will present our methodology, results, and discussion of our findings. Finally, we will conclude the paper with a summary of our results and recommendations for future research.

2 Background

2.1 Cyber-Physical System

A Cyber-Physical System is a type of system that integrates physical and computational components in a seamless manner. The physical components of CPS are connected to digital components, such as sensors and actuators, which allow them to gather and use data to control the physical components. CPSs can range from small, simple systems to large, complex systems that operate in real-time, with tight coordination between the physical and computational components. The goal of CPSs is to bring together the best of both worlds: the physical world's ability to interact with the environment, and the digital world's ability to process information quickly and accurately. This results in systems that are capable of making real-time decisions, reacting to changes in the environment, and providing new levels of efficiency, safety, and security [3–6].

The components of a CPS can be classified into two categories: physical components and computational components. The physical components, such as sensors and actuators, serve as the interface between the physical and computational domains, enabling data acquisition and physical control. Sensors gather data from the environment, while actuators perform actions based on commands received from the computational components. Embedded systems, which are integrated into the physical components, control the behavior of the physical components. The computational components, on the other hand, consist of computational hardware and software systems that process and analyze data from the physical components and generate control signals for the actuators. These computational components include data storage and management systems, which store and manage the data generated by the sensors and computational components, and user interface systems, which provide a means for users to interact with the CPS and monitor its behavior [4–7].

The networking infrastructure, which includes communication devices and protocols, is also an essential component of a CPS, connecting the physical and computational components and enabling real-time data transfer and control. The components of a CPS work together to provide a seamless integration of physical and computational systems, allowing the CPS to make real-time decisions and control the physical components in response to changes in the environment [5–8]].

Communication devices can include wired or wireless technologies, such as the Internet, Ethernet, Wi-Fi, Zigbee, and other communication technologies. The choice of communication device depends on the requirements of the CPS, such as the type of data being transferred, the distance between the components, and the need for real-time communication. Communication protocols, on the other hand, define the rules and procedures for exchanging information between components. Examples of commonly used communication protocols in CPSs include the Transmission Control Protocol/Internet Protocol (TCP/IP), User Datagram Protocol (UDP), and Real-Time Transport Protocol (RTP). In addition, there are several industrial communication protocols that are specifically designed for use in CPSs, such as the Modbus protocol for industrial control systems and the DDS protocol for real-time systems. These protocols provide features such as reliable data transmission, real-time data transfer, and security, which are critical for the functioning of CPSs. It's worth noting that communication devices and protocols are also critical from a security perspective, as they provide a potential attack vector for malicious actors. To address this, CPSs often use secure communication protocols, such as encrypted communication, secure authentication, and access control mechanisms to ensure that sensitive data and control signals are protected.

2.2 Data Transmission

Data transmission is the process of transferring data from one device or system to another over a communication channel or network. There are several different methods of data transmission that can be used, depending on the specific requirements of the communication system and the types of data being transferred. One of the most common methods of data transmission is serial communication, in which data is transmitted one bit at a time over a single communication channel. This method is often used in simple systems or for short-range communication, such as between a computer and a peripheral device. Another method of data transmission is parallel communication, in which multiple bits are transmitted simultaneously over multiple communication channels. This method is often used for high-speed data transfer, such as between a computer and a printer or scanner. Digital transmission is another type of data transmission that involves encoding data in binary form and transmitting it as electrical signals or light pulses over a communication channel. This method is often used for long-range communication, such as in telecommunications networks or satellite communication. Analog transmission is a method of data transmission that involves converting digital data into analog signals, which are then transmitted over a communication channel. This method is often used for voice communication or for transmitting data over analog communication channels, such as radio or television broadcasts. The choice of data transmission method depends on factors such as the nature of the data being transmitted, the distance over which the data must be transmitted, and the speed and reliability requirements of the communication system. In the context of CPS, data transmission over the internet plays a critical role in enabling real-time monitoring and control of physical systems. By transmitting sensor data and control signals over the internet, CPS can be used to automate and optimize complex industrial processes, transportation systems, and other physical systems. A data transmision system in Fig. 1 is showing data generation, processing and transmision over communication channel from server (control system) to the clients (robots).

There are several different types of internet connections available, each with its own benefits and drawbacks. Some of the most common types of internet connections are:

- ADSL (Asymmetric Digital Subscriber Line): This type of internet connection uses traditional copper telephone lines to transmit data. ADSL connections are often slower than other types of connections, but they are widely available and more affordable.
- Fiber: Fiber internet uses fiber-optic cables to transmit data, which results in faster and more reliable internet speeds than ADSL. Fiber connections are typically more expensive, but they are becoming more widely available in many urban areas.
- 3G (Third Generation): 3G is a mobile internet technology that allows users to connect to the internet using their mobile devices, such as smartphones and tablets. 3G connections are slower than fiber or 4G connections, but they are

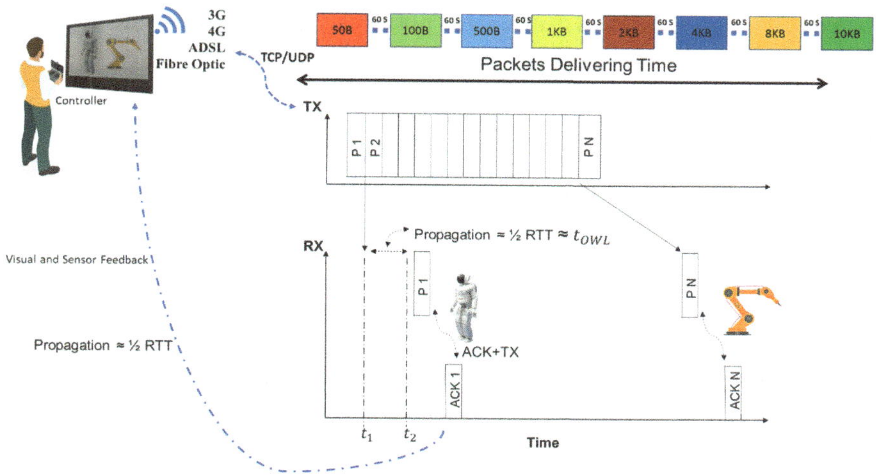

Fig. 1 Data generation, processing and transmission for CPS

widely available and can provide internet access in areas where other types of connections may not be available.

- 4G (Fourth Generation): 4G is the mobile internet technology and provides faster internet speeds than 3G. 4G connections are widely available in many urban areas, and they are ideal for streaming video and other high-bandwidth activities.
- Wi-Fi is a wireless networking technology that allows devices to connect to the internet and communicate with each other over a wireless network. Wi-Fi uses radio waves to transmit data between devices, such as smartphones, laptops, tablets, and the internet. Wi-Fi networks are typically set up using a router, which serves as a central hub for transmitting and receiving data to and from the internet and other devices. Wi-Fi is a popular technology for providing wireless internet access in homes, businesses, and public areas. It allows users to connect to the internet without the need for physical cables, which provides greater flexibility and mobility. Wi-Fi networks can be secured using various encryption protocols to protect the network from unauthorized access and maintain the privacy and security of the data being transmitted.

Measuring different internet connections typically involves using specialized software tools to collect data on various performance metrics, such as latency, packet loss, and bandwidth. Latency refers to the amount of time it takes for a system to respond to a user's request or input. In computing, latency can refer to the delay that occurs between a user entering a command or request and the system's response to that command. Latency can be caused by a variety of factors, including network congestion, hardware limitations, and software issues. Besides, reliability refers to the consistency and stability of a system's performance over time. A reliable system is one that consistently functions as expected, without errors or unexpected downtime. Reliability can be impacted by a variety of factors, including hardware quality, software stability, and system maintenance practices. Reliable systems are important in a variety of applications, particularly those that involve critical functions such as financial transactions or medical equipment.

Latency in internet communication can be measured or predicted in several ways. Here are some common methods:

- Ping: Ping is a commonly used tool to measure latency in internet communication. It sends a small data packet to a destination server and measures the time it takes to receive a response. This can give you a good idea of the latency between your device and the destination server.
- Traceroute: Traceroute is another tool that can help measure latency. It sends a series of packets to the destination server, and records the time it takes for each packet to travel through each router along the way. This can help identify any bottlenecks or slow points in the network that may be contributing to latency.
- Network monitoring tools: There are a variety of network monitoring tools that can be used to measure latency, such as Wireshark, Nagios, or PRTG Network Monitor. These tools can help identify specific areas of the network that may be contributing to latency and provide more detailed information about the performance of the network.
- Prediction based on distance: The speed of light is a fundamental limitation on the speed at which data can travel over a network. As a result, you can estimate latency based on the physical distance between two points. For example, the latency between two points on opposite coasts of the United States will be higher than the latency between two points within the same city.
- Prediction based on network congestion: Network congestion can cause increased latency, as packets are delayed while waiting for available network resources. Predicting potential network congestion can be done through various monitoring tools and can allow to predict possible increase of latency.

Here are some common methods to measure or predict reliability in communication systems:

- Mean Time Between Failures (MTBF): MTBF is a common reliability metric that measures the average time between system failures. This metric is useful for predicting how frequently a system is likely to fail and can help inform decisions about maintenance and replacement schedules.
- Failure Mode and Effects Analysis (FMEA): FMEA is a technique used to identify and analyze potential system failures and their consequences. This method involves identifying failure modes, determining their likelihood and severity, and developing mitigation strategies to reduce the risk of failure.
- Redundancy: Redundancy is a design approach that involves creating duplicate or backup components to provide fault tolerance and increase system reliability. Redundancy can be used at various levels of a communication system, from hardware to software to network topology.
- Quality of Service (QoS) Monitoring: QoS monitoring involves measuring the performance of a communication system against a set of predefined standards or

requirements. This can help identify issues that may impact reliability, such as network congestion, bandwidth limitations, or packet loss.

- Service Level Agreements (SLAs): SLAs are contractual agreements between service providers and their customers that define the level of service to be provided, including reliability guarantees. SLAs can provide a clear understanding of the expected level of reliability and define the consequences if service levels are not met.
- Latency and reliability are the main metrics for evaluating the performance of internet connections. Latency is the time required for a packet to travel across a network. It may be measured by Round-trip time (RTT) or One Way Latency (OWL). RTT is the time it takes a given network packet to travel from source to destination, and back. RTT is an important metric to determine the overall health of a network connection, and it has been used to diagnose the speed and reliability of network connections. Parameters such as network topology, bandwidth, the size of data packets, queuing, protocols, and network user traffic can impact RTT. OWL can be estimated by dividing the round-trip time in half [9]. This approximation is reasonable if the systems used in the test are symmetrical, meaning they are running the same software, using the same settings, and they have an equal network and system performance. According to Fig. 1 the One-way latency can be calculated as following equation:

$$t_{OWL} = t_2 - t_1 \quad (1)$$

Equation 1 can be used for estimating latency, t_{OWL} is the one-way latency, t_2 is the time for receiving the data packet transmission and t_1 is the time of transmitting the data packet. The latency may be impacted by any element in the chain which is used to transmit and receive data.

Reliability in communication systems could be interpreted as the capability to deliver a data packet within a certain time by the system. The concept of reliability versus latency in communication systems is shown in Fig. 2 [10, 11].

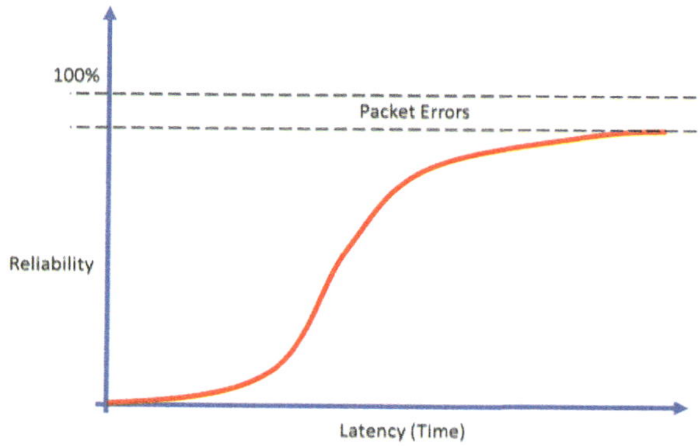

Fig. 2 Conceptual illustration of latency-reliability function [10, 11]

In this figure cumulative reliability probability in percent is shown on the y-axis versus a particular latency value (in seconds) on the x-axis. The graph starts with a reliability of 0% for 0 s delay and will increase to a maximum reliability percentage corresponding to the worst-case latency value that was experienced. Two parameters can change the shape of the curve in that figure [9, 12]:

- The variability in the delay of transmission between two devices due to various factors, including changes in network conditions, queuing delays, and processing delays is referred to as inconsistency of latency over time. This inconsistency can have an impact on the user experience and performance of applications that depend on low-latency connections, such as online gaming and video conferencing.
- The likelihood of losing data packets or the entire connection during communication, known as the probability of loss, is impacted by several factors. These factors include communication network congestion, packet drops, interference, and failures of devices in the infrastructure.

They are four main types of wired internet connections currently being used.

- Ethernet network is a networking solution that uses coaxial and twisted pair copper cables to create Local Area Networks (LAN) and Wide Area Networks (WAN). The common data rates are 100 Mbps or 1 Gbps using twisted pair cables. Unless the network is heavily congested, the latency is very low, below or approximately 1 ms, for a typical Ethernet cable of a few hundred meters or less.
- ADSL is a technology that enhances data transmission on copper telephone lines, but due to high interference, it requires various methods to improve reliability and data rates. ADSL technology has evolved significantly, with ADSL2 providing multichannel transmission that applies different latency characteristics to each channel and supports data rates up to 12 Mbps. A typical ADSL connection has a latency range of 10–20 ms, but congestion can result in a poor experience for end-users due to co-channel interference.
- Optical Fiber is the primary high-speed network that provides high-speed Internet connectivity using high-speed optical fibers. However, the latency of optical fiber connections can be several milliseconds due to the limitation of the speed of light, especially if the data source and destination are physically distant. In cases where there are multiple hops or transmission through several routers, the latency can increase to tens of milliseconds, depending on the device connected to the last-mile.
- Cable Internet services offer higher bandwidth compared to DSL through the use of optical fiber connections to the street level and co-axial cables to individual premises. Cable technology currently supports data bandwidths of up to around 300 Mbps in many areas, while most forms of DSL reach speeds of only up to 100 Mbps. The typical latency of a cable modem is between 14 and 18 ms.
- A cellular network or mobile network is another commonly used communication network where the link to and from end nodes is wireless. The network is distributed over land areas called "cells", each served by at least one fixed-location transceiver. Nowadays 3G mobile networks can provide a data rate in the

range of 1.5–6.8 Mbps by High-Speed Data Packet Access (HSPDA) protocol and 4G or Long Term Evolution (LTE) cellular networks have lower latency, better coverage range, and security properties and finally higher real-world data rates from 12 Mbps up to 22 Mbps. Latency can vary for different cellular communication networks from tens of milliseconds for LTE to hundreds of milliseconds for GPRS technology. Besides, Wi-Fi is a wireless networking technology that allows devices such as computers, mobile devices, and other equipment to interface with the internet. It allows these devices-and many more- to exchange information with one another, creating a network. This technology works at radio frequencies of 2.4 GHz or 5 GHz and the accessible data rates now can reach 500 Mbps by using multiple antenna technology while new versions of the Wi-Fi protocol will provide high data rates of several Gbps. The typical latency of a public Wi-Fi network is in the range of a few tens of milliseconds, but Wi-Fi networks are subject to self-interference with multiple access points sharing the same unlicensed spectrum.

The latency distribution in all internet connections is affected by communication delays and parameters such as:

- the type of traffic being carried and the number of users sharing the channel;
- queuing delays;
- number of routers used to connect the communicating devices;
- the bandwidth of the communications links in use;
- the data packet size;
- data transmission protocols in use.

3 Method

Previously latency performance based on ping packet has been studied in [10]. Later, using the developed test-bed created in [13], the trade-off between different data packet sizes and end-to-end latency achieved using TCP and UDP was studied. The novelty of the approach is based on the implementation of highly accurate GPS-based time synchronization in the test-bed, which enabled the measurement of the performance of off-the-shelf communication technologies. The study provides a comprehensive practical evaluation of the typical United Kingdom internet performance for internet-based cyber-physical system applications. Multiple clients were deployed simultaneously using different communication technologies to assess the performance of multiple last-mile link types. The clients communicated with a server, which used a fiber broadband connection, and data transmissions of different data packet sizes from 50 bytes to 10 Kbytes were transmitted using both TCP and UDP protocols, and the results were recorded. The implemented algorithm both for server and clients are shown in Fig. 3.

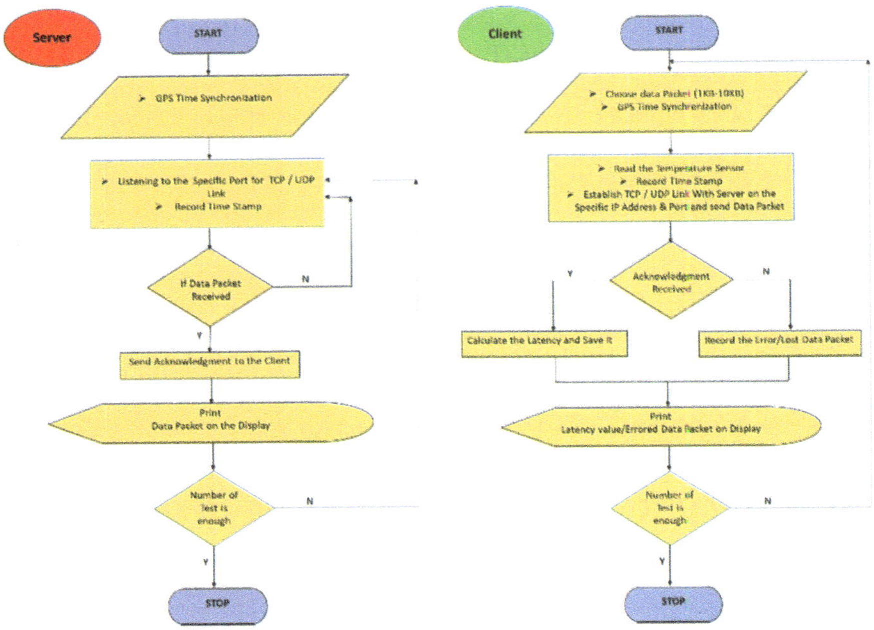

Fig. 3 Implemented algorithm in server (controller) and clients (robots) to measure communication networks characteristics

4 Results and Discussion

The presented text reports the results of an investigation into the performance of various last-mile communication technologies for TCP and UDP protocols using different data packet sizes. The investigation involved both wireless and wired connections, and the results were analyzed using Cumulative Distribution Function (CDF) plots [14], which display TCP and UDP latencies and packet loss ratios.

Table 1 displays the median latency values obtained for various data packet sizes using different last-mile communication technology solutions. The figure shows that 4G connectivity outperforms 3G connections, with a one-way latency of 242 ms for 50 byte packets and the TCP protocol. As the packet size increases to 10 kbytes, the latency slightly increases to 391 ms. For the UDP protocol, 4G exhibits slightly lower latency values of 200 ms and 324 ms for 50 byte and 10 kbyte packets, respectively. Additionally, the Table 1 highlights the poorer TCP and UDP performance of 3G for Robotic applications, with the average latency increasing from 593 ms for 50 byte data packets to approximately 1560 ms for 10 kbyte UDP data packets.

In Table 1, it is shown that wired connections exhibit superior performance for TCP. For short control packets with a size of 50 bytes, Fibre and ADSL connections demonstrate latency results of 97 ms and 254 ms, respectively. Similar to wireless connections, as the packet size increases, the latency value also increases slowly. For

Table 1 TCP and UDP median latency for all last mile technologies

Packet size	ADSL		Fibre		4G		3G	
	TCP	UDP	TCP	UDP	TCP	UDP	TCP	UDP
10 KB	250	320	330	290	395	300	1990	1560
8 KB	260	395	300	280	375	285	1900	1500
4 KB	170	130	270	250	355	300	1820	1490
2 KB	150	230	260	200	308	255	1750	1840
1 KB	190	175	250	195	290	250	1609	1800
500 B	180	104	250	190	275	245	1580	1750
100 B	150	100	145	155	230	195	1180	710
50 B	260	40	100	100	234	200	1020	593

the largest data packets with a size of 10 kbytes, a latency result of approximately 315 ms is observed. The corresponding results demonstrate that the median value of UDP latency is lower than TCP latency for wired connections. Although the UDP results exhibit more fluctuations for different data packet sizes, the latency generally increases with larger packet sizes. The ADSL network has close UDP results to Fibre, with latencies of 40–320 ms for 50 byte–10 kbyte packets.

Cumulative Distribution Function (CDF) plots were generated for all the data collected from the four last-mile communication solutions investigated in [13, 14]. The results consist of TCP and UDP latencies, which were determined using Eq. 1 for various packet sizes, as well as the packet loss ratio, which was calculated using Eq. 2 as follow:

$$Data\ Packet\ loss = \frac{Not\ recicived\ data\ packets}{all\ sent\ data\ packets} \tag{2}$$

Tables 2 and 3 summarize the CDF plots [14] in the following manner: the Table 2 displays the 90% confidence for TCP latency and UDP latency, and the Table 3 depicts the percentage of non-received data packets.

The text further reports that when following the red line plotted in the CDF plot in [14] as explained for Fig. 2, it is possible to determine the 90% confidence latency value for each CDF. For the ADSL link illustrated in Table 2, the 90% latency value for TCP packets is approximately 300 ms for 50 or 100 Byte packets, but it rises to 400 ms for larger packet sizes. In the case of UDP, the latency results show some improvement for 50 Byte packets, where the 90% latency is 184 ms. However, for other packet sizes, considerably higher latency results of up to 1.2 s are observed. Generally, when comparing the TCP and UDP in Table 2, it is noticeable that the TCP typically achieve the 90% confidence latency within a shorter time window, often below 500 ms. Conversely, UDP extend over a longer time window to achieve the 90% confidence latency, which often exceeds 1 s. Table 3 shows in details the percentage of errors. The highest packet loss percentage in Table 3 is detected for UDP packets with a 3G connection, with 21% of packets being lost. Additionally,

Table 2 Last mile technologies latency performance with 90% confidence

Packet size	ADSL		Fibre		4G		3G	
	TCP	UDP	TCP	UDP	TCP	UDP	TCP	UDP
10 KB	421	891	386	1670	628	811	2466	2095
8 KB	422	1041	427	1092	590	1017	2376	2030
4 KB	323	394	378	694	485	721	2274	1922
2 KB	330	1081	301	760	475	1400	2196	1856
1 KB	305	1333	287	561	457	472	2162	1809
500 B	294	365	303	787	445	956	2155	1792
100 B	291	1185	290	1059	464	971	1590	1236
50 B	334	184	271	321	478	1540	1447	1668

Table 3 Error percentage TCP and UDP packet loss

Percentage of error %	ADSL		Fibre		4G		3G	
	TCP	UDP	TCP	UDP	TCP	UDP	TCP	UDP
10 KB	0.1	5.5	0.1	5.2	0.3	7.7	0.1	21
8 KB	0.05	5.3	0	6	0.1	5	0.2	17
4 KB	0.1	2.9	0.05	3.2	0.4	3.8	0.05	16
2 KB	0.2	3.2	0.1	2.2	0.2	4.5	0.1	12
1 KB	0.1	4.2	0.05	3.3	0.1	3	0.05	8
500 B	0.1	2	0.05	3.2	0.05	3.8	0.05	12.5
100 B	0.1	3.3	0.1	3.5	0.1	4.2	0	6.5
50 B	0.05	2.2	0.1	2.5	0.05	6.3	0	11

TCP packets are showing more reliability with less packet loss percentage for both wired and wireless last mile technologies are tested in this experimental test bed from 0% for Fiber to maximum 0.4% in 4G. Moreover, to these results, the impact of signal strength level on latency in wireless cellular network 3G and 4G has been studied in [15]. In other study, compression techniques such Lemple Zive-Welsh and Huffman algorithm have been studied to improve the latency parameter for larger data packet size extensively discussed in [16, 17].

5 Conclusions

The research study outlines the implementation of a reliable and precise test setup, which employs high-precision time synchronization and GPS receivers integrated on both server and client sides, to gather field measurements. In order to enhance security, a VPN connection is established on top of the existing public network.

The study investigates the one-way latency of control-remote client connections and finds that UDP packets experience considerably higher losses than TCP packets for both wired and wireless connections. TCP is preferred for transmitting sensor measurements and other data that require critical reliability and latency. Various latency measurements were obtained based on user data traffic on different days of the week and times of day. Latency increases were observed on weekends for wired fiber links. However, no significant correlations were found between latency behavior and different communication links. The study also investigated data compression techniques, which yielded only modest improvements in latency performance at the expense of additional processing time. Furthermore, the study notes that routing data packets from Scotland to the South of England results in increased latencies due to the current internet topology in the UK.

References

1. Barenji RV, Akdag Y, Yet B, Oner L (2019) Cyber-physical-based PAT (CPbPAT) framework for Pharma 4.0. Int J Pharm 567:118445
2. Hariry RE, Barenji RV, Paradkar A (2020) From Industry 4.0 to Pharma 4.0. Handbook of smart materials, technologies, and devices: applications of Industry 4.0, pp 1–22
3. Saadati Z, Zeki CP, Vatankhah Barenji R (2021) On the development of blockchain-based learning management system as a metacognitive tool to support self-regulation learning in online higher education. Interact Learn Environ 1–24
4. Hariry RE, Barenji RV, Paradkar A (2022) Towards Pharma 4.0 in clinical trials: a future-orientated perspective. Drug Discovery Today 27(1):315–325
5. Vatankhah Barenji R (2022) A blockchain technology based trust system for cloud manufacturing. J Intell Manuf 33(5):1451–1465
6. Saadati Z, Barenji RV (2022) Toward Industry 5.0: cognitive cyber-physical system. In: Industry 4.0: technologies, applications, and challenges, pp 257–268. Springer Nature Singapore, Singapore
7. Hariry RE, Barenji RV, Azizi A (2022) Toward Pharma 4.0 in drug discovery. In: Industry 4.0: technologies, applications, and challenges, pp 221–238. Springer Nature Singapore, Singapore
8. Barenji RV, Nejad MG (2022) Blockchain applications in UAV-towards Aviation 4.0. In: Intelligent and fuzzy techniques in Aviation 4.0: theory and applications, pp 411–430
9. Laner M, Svoboda P, Romirer-Maierhofer P, Nikaein N, Ricciato F, Rupp M (2012) A comparison between one-way delays in operating HSPA and LTE networks. In: Proceedings of IEEE WiOpt, pp 286–292
10. Laner M, Svoboda P, Romirer-Maierhofer P, Nikaein N, Ricciato F, Rupp M (2012) A comparison between one-way delays in operating HSPA and LTE networks. In: 2012 10th international symposium on modeling and optimization in mobile, ad hoc and wireless networks (WiOpt), pp 286–292. IEEE
11. Popovski P, Nielsen JJ, Stefanovic C, de Carvalho E, Strom E, Trillingsgaard KF, Bana A-S, Kim DM, Kotaba R, Park J, Sørensen RB (2018) Ultra-reliable low-latency communication (URLLC): principles and building blocks. IEEE Network 32(2)
12. Elmokashfi A, Kvalbein A, Xiang J, Evensen KR (2012) Characterizing delays in Norwegian 3G networks. In: Passive and active measurement: 13th international conference, PAM 2012, Vienna, Austria, March 12–14th, 2012. Proceedings 13, pp 136–146. Springer, Berlin, Heidelberg

13. Zeinali M, Thompson JS (2019) Implementation of highly accurate test-bed for practical evaluation of wired and wireless internet based smart grid communications. In: 2019 UK/China emerging technologies (UCET), pp 1–4. IEEE
14. Zeinali M, Thompson JS (2020) Comprehensive practical evaluation of wired and wireless internet based smart grid communications of UK. IET Trans Smart Grid
15. Zeinali M, et al (2020) Performance assessment of UKs 4G cellular net-work for vehicle to grid smart grid applications. In: ULMC6GN work shop of ICC20
16. Mehdi Z, Thompson JS (2021) Impact of compression and small cell deployment on NB-IoT devices coverage and energy consumption with a realistic simulation model. Sensors 21(19)
17. Mehdi Z. PhD Thesis, Signal processing and communications for smart grid. The University of Edinburgh. https://doi.org/10.7488/era/1377

The manufacturer's authorised representative in the EU is Springer Nature Customer Service Centre GmbH, Europaplatz 3, 69115 Heidelberg, Germany. If you have any concerns regarding our products, please contact ProductSafety@springernature.com

Printed and bound by CPI Group (UK) Ltd, Croydon, CR0 4YY
30/09/2026
02225303-0002